Matthias Hofmann
Scientific Data: A 50 Steps Guide using Python

Also of Interest

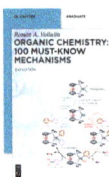

Organic Chemistry: 100 Must-Know Mechanisms
Roman Valiulin, 2023
ISBN 978-3-11-078682-8, e-ISBN 978-3-11-078683-5

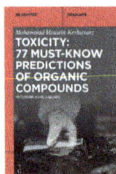

Toxicity: 77 Must-Know Predictions of Organic Compounds
Including Ionic Liquids
Mohammad Hossein Keshavarz, 2023
ISBN 978-3-11-118912-3, e-ISBN 978-3-11-119092-1

Data Management for Natural Scientists
A Practical Guide to Data Extraction and Storage Using Python
Matthias Hofmann, 2023
ISBN 978-3-11-078840-2, e-ISBN 978-3-11-078843-3

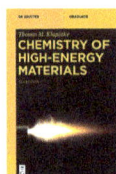

Chemistry of High-Energy Materials
Thomas M. Klapötke, 2022
ISBN 978-3-11-073949-7, e-ISBN 978-3-11-073950-3

Matthias Hofmann

Scientific Data: A 50 Steps Guide using Python

—

DE GRUYTER

Author
Dr. Matthias Hofmann
Peter-und-Paul-Weg 1
84558 Tyrlaching
Germany
matthias.j.hofmann@gmx.de

ISBN 978-3-11-133457-8
e-ISBN (PDF) 978-3-11-133460-8
e-ISBN (EPUB) 978-3-11-133470-7

Library of Congress Control Number: 2024938451

Bibliographic information published by the Deutsche Nationalbibliothek
The Deutsche Nationalbibliothek lists this publication in the Deutsche Nationalbibliografie;
detailed bibliographic data are available on the Internet at http://dnb.dnb.de.

In eternal memory of my mother, **Anita** Hofmann,

who taught me the beauty of kindness by example. This book is a tribute to your generous spirit and the profound impact you had on all of us. May your legacy of love and compassion be an inspiration to many.

Acknowledgements

Writing this book would not have been possible without the help and support of people I had the pleasure to work with in both academia and the chemical industry.

I would also like to express my gratitude to the team at De Gruyter for allowing me to write this book as an extension to *Data Management for Natural Scientists – A Practical Guide to Data Extraction and Storage Using Python* with the intention of serving R&D scientists improve their data handling, analysis and visualization capabilities. More than ever before, I am convinced that these skills need to be considered holistically, and are essential to the skill set of any modern era practitioner in the scientific field.

Finally and endlessly, my gratitude goes to Katharina.

Mission

The book is intended to serve as a "hands-on" guide to lower-level data processing routines for data generated in the context of natural sciences such as chemistry or materials science.

Referring to the Data – Information – Knowledge – Wisdom (DIKW) pyramid, the target here is to provide recipes for systematically ascending the steps of this model to speed up and improve the quality of your research. Everything shown here merely serves as a tool to support frequently encountered challenges in scientific workflows. Therefore, the focus is on the high-level programming language Python for all the concepts described here.

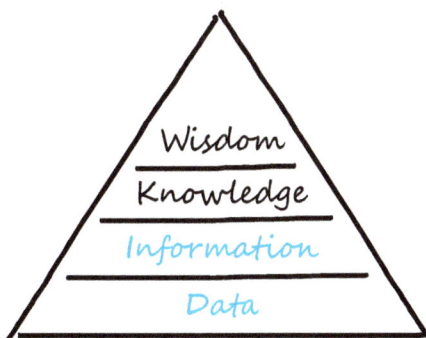

The hypothetical scenario of this book is setting up a research project from scratch. In addition to the physical preparations needed, the following concepts recommend steps to additionally set up your project for success from a data management, analysis and visualization point of view. The process is illustrated with the help of a simplified real-world example.

https://doi.org/10.1515/9783111334608-201

Scientific Data: A 50 Steps Guide using Python provides a compilation of the data processing steps that are essential for natural scientists. It aims to point out the dualism of classical natural sciences as taught in universities and the ever-growing need for technological/digital capabilities. The book covers a representative use case where both technical issues and scientific aspects are equally considered. After the initialization of a sustainable Python environment, the discussion shifts from these technical aspects to more scientific issues such as the extraction of meaningful characteristics and multi-objective optimization. The concise problem-solution-discussion structure throughout the chapters, supported by Python code snippets, emphasizes the book's ambition to serve as a practitioner's guide. It will give you a solid understanding of how to analyze experimental data in a common natural scientific context and how to ensure the sustainable use of your findings.

Audience

The book is addressed to graduate students of the natural sciences willing to improve their data handling skills and capabilities for long term accessibility of their experimental results. Another potential audience for the book are university professors who want to ensure a sustainable long-term use of experimentally collected data over several "generations" of students. An additional possible target audience could be small or medium-sized companies in the described field that are unable or unwilling to employ a full-time developer to handle advanced database environments and scientific staff with a strong background in data management. In this sense, the goal of the book is to promote a mutual understanding of the needs of both scientists and data engineers.

The greatest value of this book is probably inherent for (self-considered) *science dataists*. Although I could not find a strict definition of this term,[1] it refers to my understanding of a (trained) R&D scientist using *data science* tools as an "add-on" to improve the speed and quality of his or her natural scientific work.

Assumptions

This book does not require in-depth previous knowledge of Python or data management. Some experience with scripting languages is certainly helpful but not mandatory for understanding the key ideas of this book and along with the code snippets provided.

The examples presented here were created using Python 3 on a Windows machine.

---

1 ... to the best of my knowledge.

This book is not ...

This book is not intended as a full-stack reference for all of the *packages* and functions used for the purpose of organizing the data related to the exemplary scientific question. Hence, there will be no detailed explanations of *arguments* and *keyword arguments*, i. e., the arguments passed to a Python function. This is particularly the case for those arguments, which are not used in the context of the examples presented. At this point, reference is made to the extensive and up-to-date online documentation of the relevant packages. For these sites, the documentation appropriate for your installed version is readily available.[2,3,4] Furthermore, this book does not claim to show programming excellence in every line of code or syntactic perfection. Rather, it condenses the experience I have gained over the last few years working with Python in a natural scientific context on a daily basis. In short: This is a book by a so-called *practitioner* for aspiring practitioners.

Contents of this book

Basics

Summarizes the first steps in setting up your R&D project for success. In particular, it covers recommendations for getting Python up and running, managing Python packages via virtual environments, configuring the spyder integrated development environment (IDE) and the benefits of creating a GitHub account and repositories are covered therein. It also covers synchronization via *Github Desktop* and the basics of *Markdown*.

Organization

Before we begin the deep dive, we will first review the overall concept sketch that we need to keep in mind throughout the process in order to not lose sight of our target. Next, we will demonstrate initializing a project via the `poetry` package and tracking the environment created. In addition, `poetry` is also helpful for sharing projects once they are complete (or in an intermediate state). The remaining concepts in this section cover path handling on local drives, the benefits of writing convenience functions, the use of *toml* files and introduce the concept of *testing*.

---

2 https://pandas.pydata.org/docs/
3 https://seaborn.pydata.org/api.html
4 https://matplotlib.org/stable/api/index

Interfacing with common data formats

Handling experimental results or other source files is an integral part of hands-on scientific R&D work. Accordingly, reading the contents of Microsoft® Excel®, *txt*-, *Microsoft Word®* - and *pdf* -files. Retrieving information from websites is also briefly covered. The section concludes with a short excursion to *regular expressions* and reading/writing to a SQLite database.

Planning experiments and/or building on legacy data/information

This is where the blank sheet approach meets the reality of legacy data. This section presents both concepts for proposing meaningful experiments from scratch and for leveraging already existing experimental results and conditions.

Collecting experimental data / lab work phase

Here we describe approaches for collecting experimental data in an efficient way during the lab work phase. In the case of the first approach of using dedicated Python packages that are already available, we move on to towards the building the missing parts[5] (in our opinion).

Visualization of experimental results

For the visualization of experimental results, plotting using the packages `matplotlib`, `seaborn` and `plotly` is demonstrated. Additionally, some concepts for visualizing multidimensional datasets are presented.

Approaching the scientific questions (modeling and recommendation)

This section covers the heart of the R&D process, which benefits most from a combination of in-depth scientific knowledge and advanced data processing and analysis skills. This is where the previously mentioned *science dataists* tend to shine.

A scientific background is certainly helpful for selecting *relevant* data and information. However, modeling the relations between defined inputs and obtained outputs is more in the realm of data science. With the models as – ideally – *digital twins* available

---

5 Ideally, those developments should be made available to a broader audience, if possible.

through different tools, R&D scientists can simulate and thereby validate system behavior. This clearly shows the close connection between *natural sciences* and *data science* at the very core of the *data scientific* process.

Additionally, a method is proposed for dealing with too few experiments.[6] The section concludes with the idea of numerically solving the "reverse design problem" applying multi-objective optimization and an introduction to formalized causal inference.

Sharing the project

Ideally, sharing the results of your work concludes a hopefully successful project. Once again, the poetry package is shown to be an effective tool for building files for distribution and publishing the package to package indices such as The Python Package Index (PyPI). Additionally, the option of sharing contents via streamlit applications is demonstrated.

Further reading

This concluding section addresses additional topics such as ensuring *code styling* using the black package, configuring pre-commit and a primer on building standalone solutions using the PyQt package.

Conventions used in this book

The following typographical conventions are used in this book:

Italic
for new terms, URLs, file names, file extensions, path names and directories.

Typewriter
for commands, options, variables, attributes, keys, functions, types, classes, methods and modules.

SMALL CAPS
for Structured Query Language (SQL) keywords and queries.

This environment indicates additional information.   i

------

6 As many of experiments are costly and/or time consuming, handling with comparably small datasets is quite common.

This environment indicates that attention should be paid to the issue described.

This environment indicates the option for some exercise based on the mentioned code snippet.

Concerning code snippets

The attribution of code snippets considered useful for your project is appreciated but not necessarily required. An attribution regularly includes the title, author, publisher and ISBN. For the present case, this could be *"Scientific Data: A 50 Steps Guide using Python* by Matthias Hofmann, 2024, De Gruyter, ISBN-978-3-11-133457-8".

The exemplary experimental results files and Python scripts shown in the following are available free of charge on
–   https://www.degruyter.com/document/isbn/9783111334608/html, and
–   https://github.com/mj-hofmann/Concepts-Use-Case.

Microsoft® copyrighted content

The references to and screenshots of Microsoft products herein comply with the conditions of Use of Microsoft copyrighted content (https://www.microsoft.com/en-us/legal/intellectualproperty/copyright/permissions). Accordingly, they are used with permission from Microsoft.

Contents

Introduction and challenge

In the contemporary scientific landscape, the importance and impact of data science across disciplines cannot be overstated. This trend is particularly relevant in fields such as chemistry and materials science, where the generation and analysis of experimental data is fundamental to research endeavors. However, despite the increasing availability of data and the potential insights it holds, the education of natural scientists in these fields often falls short of equipping them with the necessary data (science) skills.

Traditionally, the emphasis in the education of chemists and material scientists has been on theoretical knowledge and experimental techniques. While proficiency in conducting experiments is undoubtedly essential, the ability to effectively analyze and derive meaningful insights from the resulting data is equally crucial. Unfortunately, many scientists in these fields lack the training and expertise necessary to harness the full potential of modern data analysis techniques.

By integrating data skills into the training of natural scientists, we can unlock numerous benefits for research quality and innovation. For instance, data science techniques such as machine learning algorithms have demonstrated immense potential in creating new materials with desired properties or optimizing chemical processes. These advancements not only accelerate the pace of discovery but also enable researchers to tackle increasingly complex challenges in their respective fields.

Moreover, proficiency in data analysis empowers scientists to extract valuable insights from large datasets, identifying patterns, trends and correlations that may not be apparent using traditional methods. This enhanced analytical capability not only strengthens the rigor and validity of scientific findings but also promotes reproducibility and transparency in research practices – both in academic and industrial research environments.

Bridging the gap between traditional scientific training and the demands of modern data-driven research requires a concerted effort in education and professional development. Incorporating routines focused on data analysis techniques and best practices for managing and analyzing scientific data into existing chemistry and materials science programs is essential.

Ultimately, equipping natural scientists with robust data science skills can greatly enhance the quality, impact, and relevance of their research efforts. Moreover, fostering interdisciplinary collaboration between chemists/materials scientists and data scientists will further catalyze innovation and drive progress in addressing pressing scientific challenges. Embracing data science as an integral part of scientific training is not only beneficial, but imperative to navigate the complexities of the modern scientific landscape.

https://doi.org/10.1515/9783111334608-001

Basics

The concepts introduced in the next pages will serve as the basis for the considerations that follow. Please, keep in mind that a highly *personally opinionated* approach is used here.

It aims to cover topics that will assists in keeping your code base easier to maintain and share in the long run. A schematic representation of *why* you should opt for the approach presented here, which includes the use of a dedicated version control system, is given in Figure 1.

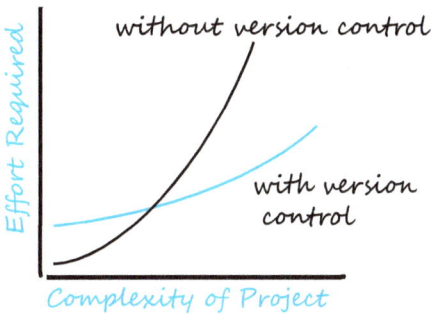

Figure 1: Evolution of required effort and time for maintaining projects of increasing complexity with and without version control setup compared.

By skipping the initial effort of setting up version control and a dedicated environment, you can get started faster, but you can also reach the limits (in terms of time required) at considerably lower levels of complexity.

https://doi.org/10.1515/9783111334608-002

1 Getting hands on Python

Problem

You want to install Python and regularly use packages on your machine.

Context

If you are starting from scratch on your machine, you will need an installation of Python and some additional packages for, e. g., visualization and analysis that do not come with Python alone. *Anaconda* has proven to be a reliable so called *distribution* for simplified package management and deployment.

Solution

Download the installer (for Windows or Mac) from https://www.anaconda.com/ and follow the instructions of the installation wizard.

To verify the installation status, open the *Anaconda Prompt* and enter the commands `conda --version` and `python --version` to get the installed versions as indicated in Figure 1.1.

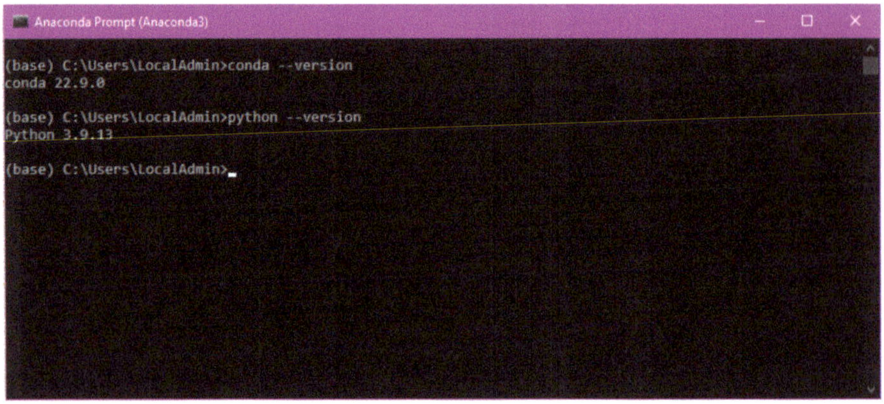

Figure 1.1: Verifying the installations of Anaconda and Python via the Anaconda Prompt.

https://doi.org/10.1515/9783111334608-003

Discussion

In addition to providing the distribution, Anaconda is also a valuable resource for further information through newsletters and the periodic *State of Data Science report*,[1] which highlights trends in the industry.

Further reading

If you are having trouble with verifying your installation, consult the Anaconda troubleshooting guide (https://docs.anaconda.com/reference/troubleshooting/) for possible sources of error. Also, running the Anaconda Prompt as an admin (if you have such privileges) may help in some cases.

---

1 https://www.anaconda.com/state-of-data-science-report-2023

2 Using virtual environments

Problem

You want to ensure a specific combination/selection of packages in specific versions that will run with a specific version of your Python installation.

Context

Typically, there is no need to build each and every part of, e. g., a data processing pipeline from scratch. In fact, as will be shown in a later concept, we explicitly want to use as much as possible (and as much as is considered useful) of what is already available, and adapt it to our needs.

This, however comes with the drawback that we are dependent on the defined version of the packages used in our code. This is where *Virtual Environments* come in. By relying on these environments, a defined combination of Python and package versions can be bound together and compiled for distribution.

Solution

Open the *Anaconda Navigator* coming with the installation of Anaconda, navigate to the *Environments* in the left menu and click the *Create* button as shown in Figure 2.1.

In this example, we create a virtual environment named py39 using Python version 3.9.16.

Discussion

To verify the successful creation of the py39 environment, we again rely on the Anaconda Prompt. After starting this console, there will be a *(base)*-string left to the home path, indicating that we are currently using the base environment as installed with Anaconda (see Concept 1). Using the conda-commands summarized in Table 2.1, we can switch between the different environments, check the corresponding Python-versions and list the installed packages including version information using the command

```
pip list
```

A screenshot of the commands including the output is shown in Figure 2.2.

https://doi.org/10.1515/9783111334608-004

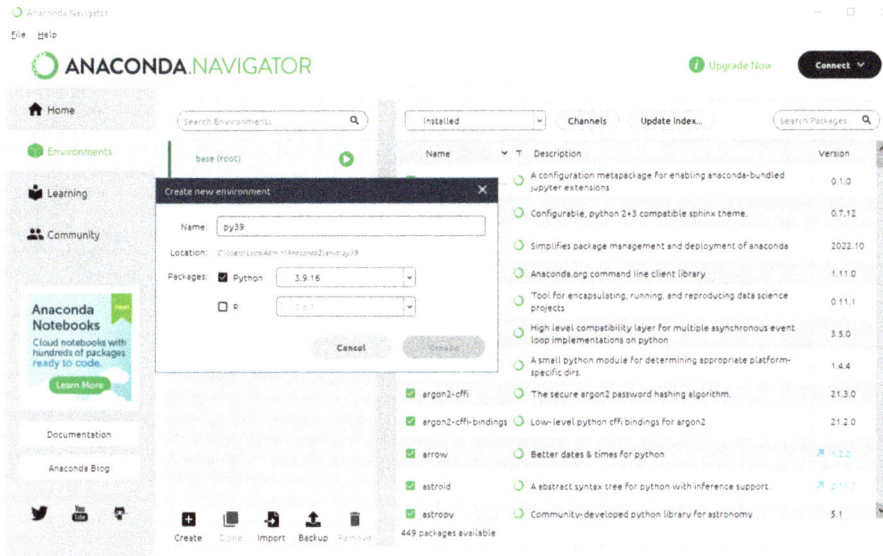

Figure 2.1: Creation of a the virtual environment py39 via Anaconda Navigator.

Table 2.1: Overview of frequently used commands for managing virtual environments using conda.

Task	Command
verify conda install and check version	conda info
activate the environment ENVNAME (before trying to install packages in this environment installing)	conda activate ENVNAME
get back to base environment	conda deactivate
list installed packages	conda list
list all environments and locations	conda env list
export environment information (platform + package specific)	conda env export ENVNAME>ENV.yml
import the enviroment ENVNAME from a *.yml* file	conda env create -n ENVNAME --file ENV.yml

Further reading

A cheat sheet listing a summary of conda commands is available online (https://docs.conda.io/projects/conda/en/latest/_downloads/843d9e0198f2a193a3484886fa28163c/conda-cheatsheet.pdf).

Additionally, the information for restoring the environment you just created can be saved to a *yml*-file to recreate it on another occasion or to share it with colleagues and/or other contributors using the command

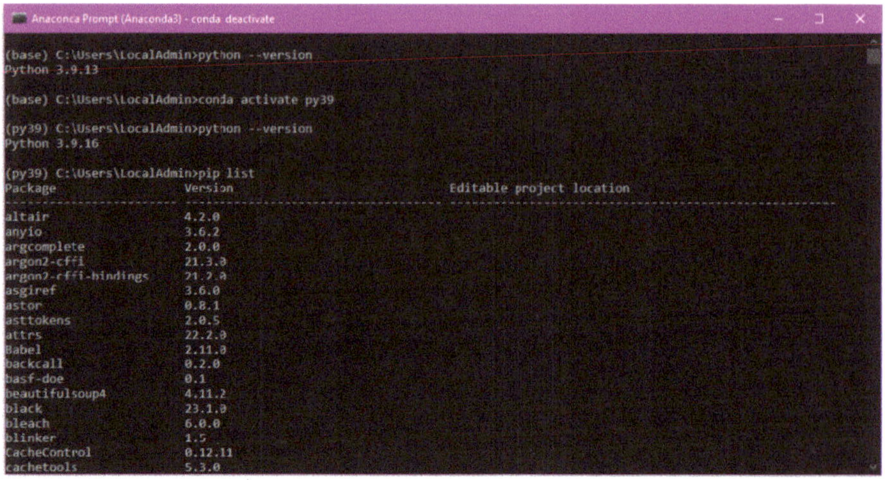

Figure 2.2: Switching from the Anaconda base environment with Python-version 3.9.13 to the previously created virtual environment py39 with Python-version 3.9.16. An overview of the installed packages is accessible via the `pip list` command.

```
conda env export >py39.yml
```

Note that exporting an environment as a yml file is also possible using the *Backup*-button in the *Environments* section of *Anaconda Navigator*. Following the commands in the cheat sheet referred to above, the appropriate environment can be recreated from the yml file.

3 Configuring your integrated development environment

Problem

You want to use the previously (see Concept 2) created virtual environment for scripting and coding in your chosen IDE.[1]

Context

The virtual environment created so far only defines a set of tools available for certain scripting tasks. The "packaging" into environments allows the code to be run at a later time as newer versions of Python (and especially third-party packages) are released. Using virtual environments also ensures smooth collaboration between you and your colleagues working on different physical machines. The present section shows how the virtual environment concepts can be used within the IDE, i. e., the program where the coding takes place.

Solution

Open the *Anaconda Prompt*, activate the environment we want our IDE to be aware of, and show its information summary. The commands and console output are shown in Figure 3.1.

In particular, we are interested in the information provided in the line *active env location*. This is the path where the environment installation of Python and its installed packages "live" are located. Next, we want to share this information with our chosen IDE, *spyder*. Therefore, we open this application, navigate to *Tools* ⇒*Preferences* ⇒*Python interpreter*. In the text entry labeled "Use the following Python interpreter" we paste the just obtained *active env location* followed by *python.exe* according to Figure 3.2.

After successfully specifying the environment, its name and the Python version is printed in the bottom line of the IDE. After a restart, an error message similar to the one shown in Figure 3.3 may be displayed in the console window. To solve this issue, run the suggested command in the *Anaconda Prompt*, wait for the installation to complete, and restart spyder.

---

1 Here we will use *spyder*, which comes with the installation of *Anaconda*.

https://doi.org/10.1515/9783111334608-005

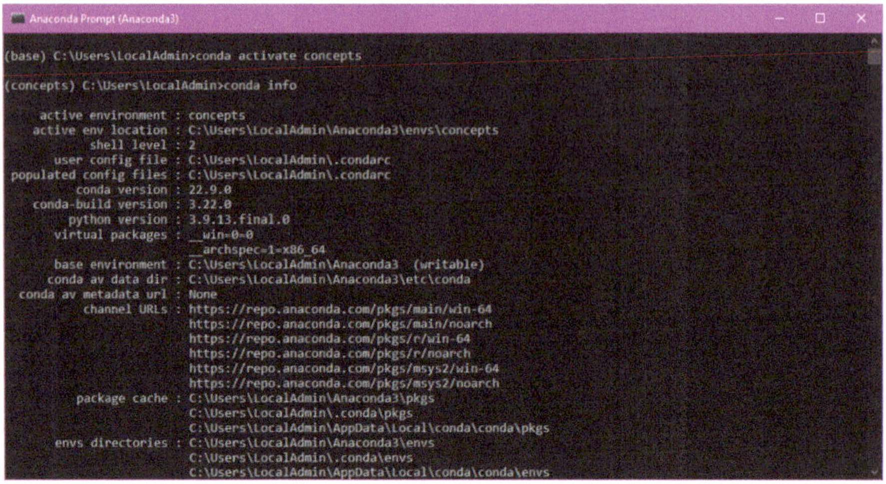

Figure 3.1: Locating the active virtual environment for further use in the IDE via the commands `conda activate concepts` and `conda info`.

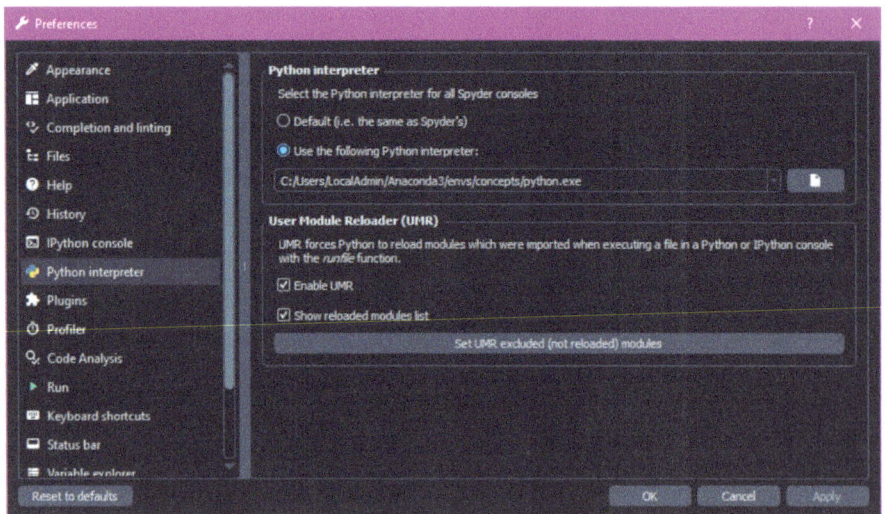

Figure 3.2: Specifying the Python executable of the virtual environment `concepts` in the spyder IDE.

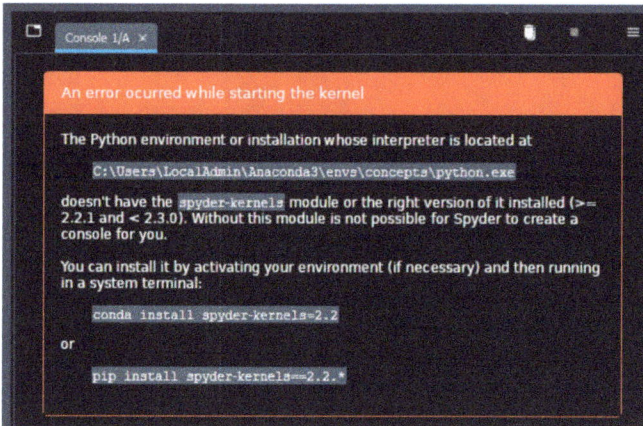

Figure 3.3: Missing `spyder-kernels` module error.

Discussion

Switching between multiple virtual environments is also possible in other popular IDEs such as Visual Studio Code (VS Code)[2] or PyCharm[3].

Further reading

For more information on handling multiple virtual environments in VSCode and Py-Charm, see the links below.
- VS Code: https://code.visualstudio.com/docs/python/environments
- PyCharm https://www.jetbrains.com/help/pycharm/creating-virtual-environment. html#python_create_virtual_env

---

2 https://code.visualstudio.com/
3 https://www.jetbrains.com/pycharm/

4 Having a GitHub account

Problem

You want to work productively and securaly both on your own and also in teams with collaborators on a shared code base. Furthermore, you want to be able to track changes to your code over time, trace errors and restore specific statuses over time.

Context

When writing new code, you typically "break things". This is true for lines you write on your own, but even more so when you work collaboratively. In this scenario, the additional challenge is how a change introduced by a contributor might affect the changes you had in mind. To keep track of all these modifications and combine them into a unified working codebase, *GitHub* has you covered. Accordingly, Git is the single most popular version control system in use today.[1]

Solution

Navigate to https://github.com/, click the *Sign Up* button and follow the guided process. Upon completion, you will be able to log in to your homepage to get an overview of your recent activities as indicated in Figure 4.1.

Discussion

Please note, that having and using a GitHub account in the field of software development/coding is as common as other types of social media such as Twitter, LinkedIn, Facebook and the like.

It is both a way of sharing your work with the world and a great place to learn from successful projects. Just browse the code snippets to find out how certain aspects have been implemented in one of the projects you are working with.

In addition to GitHub, there is also GitLab, which can be considered a "private" version of GitHub, where only a selection of people can access the source code. Therefore, GitLab is typically used in the enterprise context. GitLab has some notable key advantages over GitHub, as it provides developers with an unlimited number of private repositories that can be used with a built-in continuous integration system.[2]

---

1 https://www.bairesdev.com/blog/which-version-control-system-developers-use/
2 https://www.upgrad.com/blog/github-vs-gitlab-difference-between-github-and-gitlab/

https://doi.org/10.1515/9783111334608-006

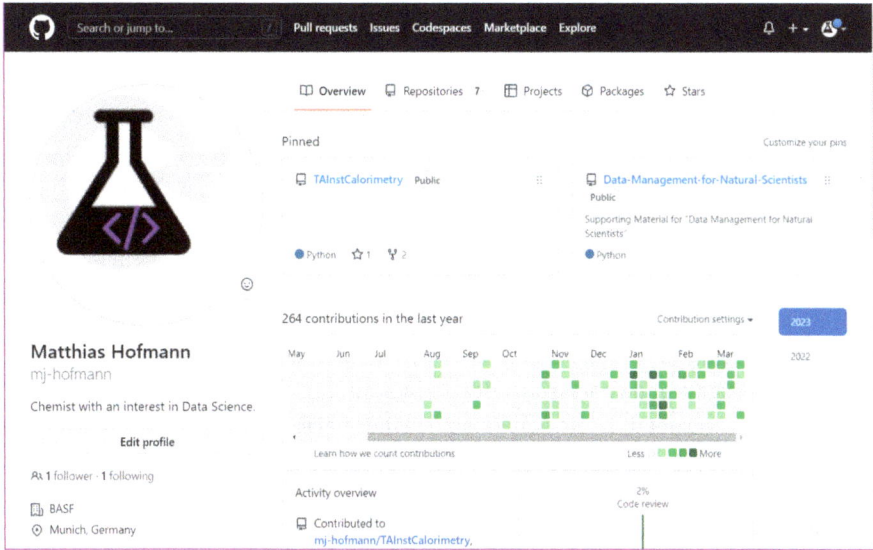

Further reading

Working with Git can be a challenge in itself and is covered in great detail in numerous books. For getting a first impression and/or a refresher on Git, check out the links below.

- GitHub quickstart: https://docs.github.com/en/get-started/quickstart/hello-world
- git cheat sheet: https://education.github.com/git-cheat-sheet-education.pdf

---

5 Creating repositories for dedicated projects

Problem

You need some "space" to store your code, supporting information, context and smaller amounts of data.[1]

Context

Assuming you want to compile a bundle of Python scripts, the packages used, experimental data, insights and conclusions drawn in a traceable manner, using a repository on GitHub (see Concept 4) is the way to go. Furthermore, the repo can be specified as *private* or *public*. This defines who can see your repo. You should also create a repo with a README file to provide room for a more extensive description, add a *.gitignore* file based on the Python template, and select a license based on your needs.[2]

After confirming the creation of the repo, we will be redirected to the homepage of the repo as shown in Figure 5.1.

Solution

Navigate to the *Repositories* section of your GitHub profile and press the *New*-button. The page that appears allows you to define a repo[3] name and set a short description.

Discussion

Note that the repo you just created already contains three files:
- *.gitignore* This text tells the version control system which files or folders to ignore in a project. Typically, temporary "helper" files are excluded. Some developers also want to exclude generated images or pdf files. Since this depends heavily on the intended use, just be aware, that this file is the place to go to have files either tracked or not.
- *LICENCE* Contains information about the license *chosen* for the project.
- *README* Should contain a description of what is covered in the repo, how to get it running, possible issues, an outline, and anything else you think might be relevant for someone else trying to use it.

---

1 GitHub blocks files larger than 100 MB.

2 https://choosealicense.com/

3 Typically, the term "repository" is abbreviated as "repo".

https://doi.org/10.1515/9783111334608-007

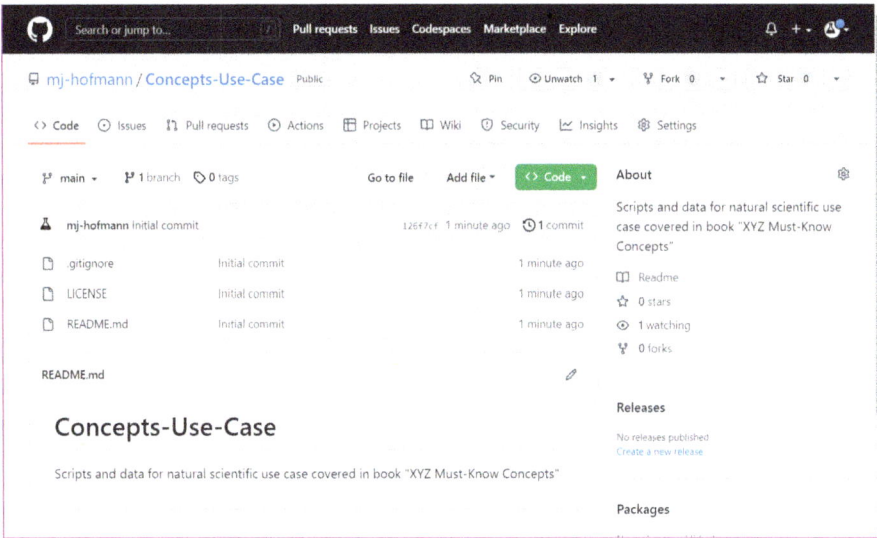

Figure 5.1: Home screen of the newly created repo *Concepts-Use-Case*.

Interestingly, the short description defined earlier has already found its way into the rendered version of the README[4] below the list of files as shown in Figure 5.1.

Further reading

- About large files on GitHub: https://docs.github.com/en/repositories/working-with-files/managing-large-files/about-large-files-on-github.
- Repository size limits for GitHub.com: https://stackoverflow.com/questions/38768454/repository-size-limits-for-github-com.
- Setting repository visibility: https://docs.github.com/en/repositories/managing-your-repositorys-settings-and-features/managing-repository-settings/setting-repository-visibility.
- Selecting a licence: https://choosealicense.com/.
- *.gitignore* How To: https://github.com/kenmueller/gitignore.

---

4 The README is formatted using Markdown (see Concept 7).

6 Synchronizing GitHub desktop

Problem

You want to work on a code base and with the files stored in a repo (see Concept 5) locally on your computer, using the version control capabilities of GitHub.

Context

Following the previous steps, we have a dedicated "space" or repo where we can store code and files *online*. In order to mirror this repo on our machine (and also to allow for collaboration with others), we want to synchronize it. The application *GitHub Desktop* is highly recommended for this purpose.

Solution

First, go to https://desktop.github.com/ to download *GitHub Desktop* and install the application by following the steps provided by the installation wizard.

Navigate to the home screen of the repository you want to work with, e. g., https://github.com/mj-hofmann/Concepts-Use-Case, press the green *Code* button and click on the text *Open with GitHub Desktop*.

To complete the synchronization, confirm by specifying where you want to have the files *locally* and press the *Clone* button as shown in Figure 6.1.

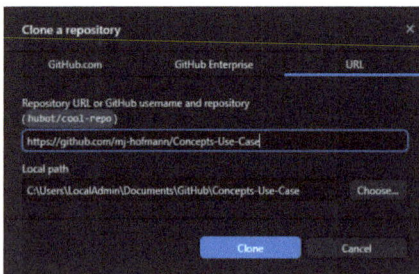

Figure 6.1: Cloning a repo to a specified local path using the "Open with GitHub Desktop" function. The action has to be confirmed by pressing the *Clone* button.

Upon successful completion, the three files mentioned in the previous concept (see Concept 5) are now available on your machine for further modification, tracking of changes, and mirroring back to the online repository (see Figure 6.2).

https://doi.org/10.1515/9783111334608-008

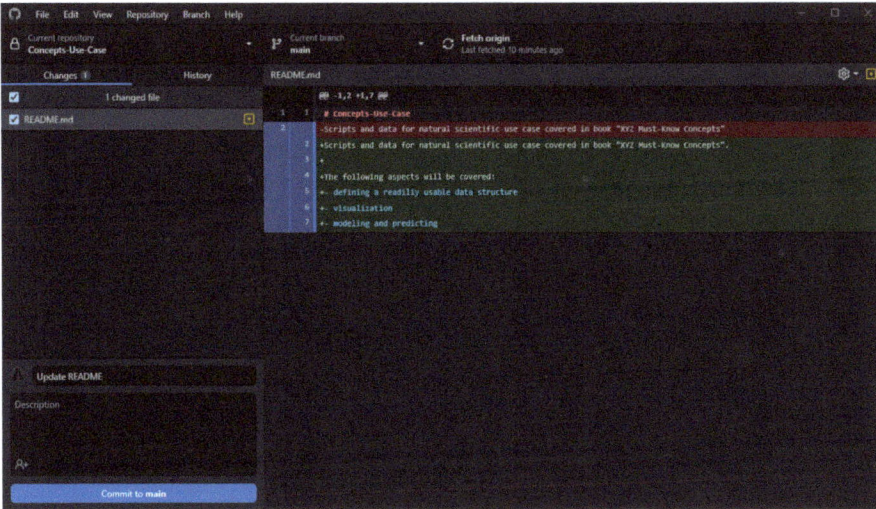

Figure 6.2: Changes in the *README.md* file are indicated in the *Changes* tab on the left hand side of the interface. The commit message "Update README" is specified for human-readable tracking of the changes. To make the changes effective on the *remote*, click the *Commit to main* and then the "Push origin" buttons.

Discussion

You may be wondering why it is important to have the same files both online and on your machine. The key idea in the scenario described here is to use the online version on *GitHub*, the so-called *remote* or *origin*, as the latest "approved" version of your code and/or project status.

Changes to your local codebase are tracked after you save the files, but are only available online after you perform the commit and push-actions. *GitHub Desktop* will take care of this for you, but of course you can also use the command line to perform these steps.[1]

Working in teams makes this approach even more interesting. Generally, developing new features, analyzing certain aspects of a dataset and compiling summaries is done locally. Once you are satisfied with the result, you mirror it to the online version. This "back and forth" between your local copy and the remote one is *the* decisive pattern of version control systems.

Accordingly, the README file has been changed to illustrate the process of making local changes available on the *remote*.

---

1 For an overview of the commands, see the reference to some cheat sheets below.

Further reading

- git – the simple guide: https://rogerdudler.github.io/git-guide/.
- git: https://git-scm.com/.
- git workflows: https://www.atlassian.com/git/tutorials/comparing-workflows.
- git cheat sheet: https://education.github.com/git-cheat-sheet-education.pdf.
- git cheat sheet (GitLab): https://about.gitlab.com/images/press/git-cheat-sheet.pdf.

7 Knowing basic markdown

Problem

You want to have a properly formatted *README*-file (see Concept 5) on the front page of your repo, including sections, lists, formatted code snippets, and images.

Context

Markdown is a lightweight markup language for creating formatted text using a text editor. It is widely used in blogging, instant messaging, online forums, collaborative software, documentation pages, and readme files.[1]

Solution

Use Markdown in your *README*-file to introduce the formatting according to Table 7.1.

Table 7.1: Overview of the most common Markdown commands.

Task	Command
Make a header	# Header
Make a sub-header	## Subheader
New paragraph	Add a blank line
Make a linebreak	Add two spaces at end of line
Add horizontal lines	---
Bold text	*text*
Italic text	_text_
Monospace text	`text`
Bullet points	- item 1
Add a link	[link](http://example.com)
Add an image	![Image](http://link-to-image.png)

Discussion

If writing Markdown is not your daily business, using an interactive editor (see list below) can be helpful both in terms of speed and achieving desired output.

Concerning the adaption of *README* files within a repo, note that they can also be edited online with the benefit of a preview tab, as shown in Figure 7.1. In order to make

---

1 https://en.wikipedia.org/wiki/Markdown

https://doi.org/10.1515/9783111334608-009

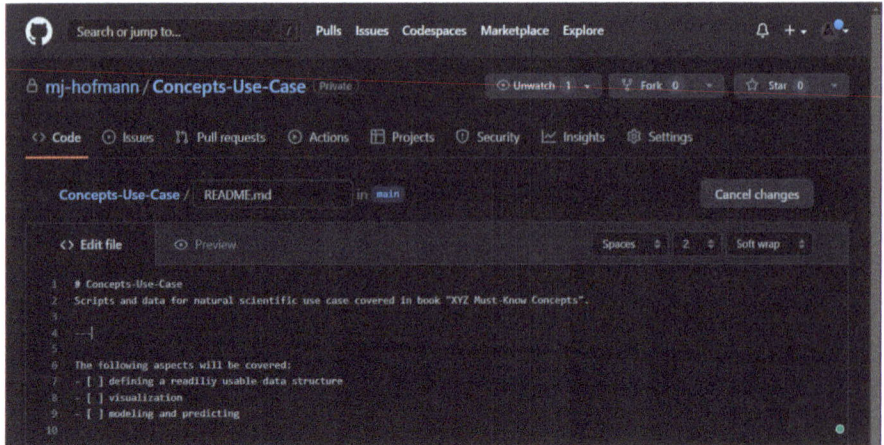

Figure 7.1: Editing a the *README* file in the online editor of *GitHub*.

these changes available to your local copy of the repo, click the "Pull origin" button in *GitHub Desktop*. The check for differences between the status on your machine and the *origin* is automatically run when GitHub Desktop starts.

Further reading

The following list summarizes some helpful online tools for editing README files using Markdown.

– StackEdit: https://stackedit.io/
– Make a README: https://www.makeareadme.com/
– Readme generator: https://readme.so/

Organization

So far, the concepts covered have focused on how to set up a Python environment for sustainable use by yourself, your colleagues, and anyone else you want to contribute.

Now, let's switch gears and introduce the natural science use case we'll be covering throughout the remainder of the book, broken down into smaller parts using the *concept* format. The goal of these *concepts* is to be sufficiently specific to highlight the challenges involved, but general or abstract enough to allow transfer to *your* scientific problem.

What is the scientific challenge?

In the following, we will consider the hypothetical case of a an adhesive formulation[1] consisting of three components A, B and C. The essential challenge of formulation is to balance out multiple factors in a possibly "ideal" way, i. e., to find an acceptable compromise of properties. In many cases, product quality – however quantified – is opposed to the formulation cost.

In our example, we want to find an *optimal* combination of the components A, B and C to yield a product with the following properties, as sketched in Figure 2:

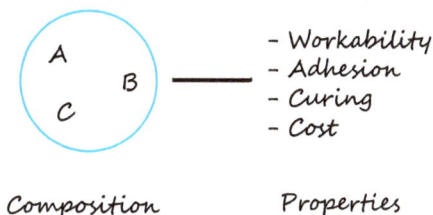

Figure 2: Conceptual relation between a formulation's composition and its resulting properties as accessible via experimental characterization application, and/or calculation.

- decent workability,
- strong adhesive power,
- short curing times and
- low cost.

Also, it is important to note that the primary task of a scientist is not to find "the optimum" as and end in itself. Rather, the task is to develop a systematic or mechanistic understanding of the relationships between the factors you can change (here, fractions of the components A, B and C) and the resulting changes in the observed, i. e., measured

---

1 https://www.acs.org/careers/chemical-sciences/fields/formulation-chemistry.html

https://doi.org/10.1515/9783111334608-010

characteristics. Once you understand (or have the ability to model) the relationships, you will be able to identify "the optimum" as a byproduct. Even more than that: You will be able to identify multiple optima related to different goals or scenarios – based on *one* dataset.

How do you actually quantify these properties?

Often, it is much more challenging to identify appropriate and reliable methods for quantifying the above mentioned characteristics. However, this should be considered the most critical point in setting up experimental plans. Only characteristics that can be determined in a reliable, repeatable and precise manner should be used in a "characterization".

Another aspect to note is *consistency*. Stick to your chosen methods, even if you find out after a certain number of experiments, that your chosen method is not absolutely perfect. The idea behind this potentially controversial approach is the following: Assuming that your method inappropriately introduces a certain bias (e. g., a miscalibrated instrument), accept this deviation and keep in mind the challenge or goal of the optimization task. As previously stated, this is usually the identification of the maximum and minimum values of certain characteristics. As sketched in Figure 3, the erroneous quantification will *not* compromise your overall conclusion. In short: *If you are doing it "wrong", make sure you are doing it consistently wrong.*[2]

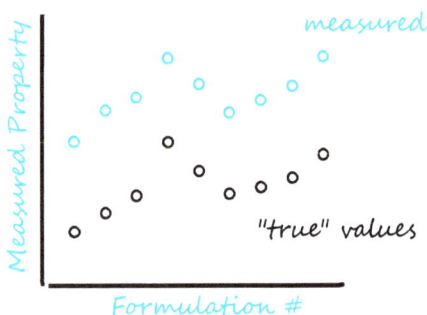

Figure 3: Because experimental error is inherent in measurements, the "true" value is all too often not immediately accessible. Staying within the limits of a characterization approach, preserving the same trend between true and measured values across samples is more relevant.

This also comes with the revelation that the collectivity of formulation experiments holds more information than a singular experiment.

---

2 In case you are uncomfortable with this approach, consider that you are often not even aware that the absolute value drawn from an experiment is actually incorrect.

For our consideration, we *choose* to use the measurable characteristics listed in Table 2 as proxies for our so far only loosely defined targets.

Table 2: Overview of measurable characteristics used as proxies for potential targets of optimization.

Target feature	Measured proxy
good workability	viscosity as measured under defined conditions
strong adhesive power	pull-off strength value
short curing times and	heat flow calorimetry parameter
low cost	sum of component fractions multiplied by unit cost

How to store and access the data?

In many cases, relying on an ordered folder structure to store experimental results and compositions is still a valid option. This comes with the benefit of considerable amounts of freedom on the one hand, but on the other hand it requires some critical thinking in terms of

- How many (sub)folders are required?
- What is the relationship between the individual files?
- Can the experiments be uniquely identified?

In the example used throughout this book, the unique identification of experiments is provided by the central file *composition.xlsx*. There are probably more sophisticated solutions than using a simple Microsoft® Excel® file to organize experiments in larger series. Nevertheless, there are certain advantages to using this approach, keeping in mind the main purpose of this type of file: to provide unique identification of experiments. Given the following easily memorable conditions, using this approach will speed up your daily processing routines:

- Each variable is stored in its own column (top down).
- Each observation or "experiment" is stored in its own row (left to right).[3]
- Do not merge cells.
- Do not use mixed value types in one column.[4]
- Include units to the column headers where applicable.
- Avoid special characters in column headers when possible.[5]

---

3 https://pandas.pydata.org/Pandas_Cheat_Sheet.pdf

4 One way to apply this database-like feature is to use data validation on cells.

5 Even today's commercially available software still suffers from inadequate handling of special characters such as parentheses, percent signs, and others.

Any other system that provides *unique* identification of sample composition, sample processing and the corresponding results is sufficient.

Given the unique identification of experiments, we only need to "attach" the measured quantities. To keep things as simple as possible, just resume the identification in the respective subfolders. An overview of the suggested folder structure is given in Figure 4.

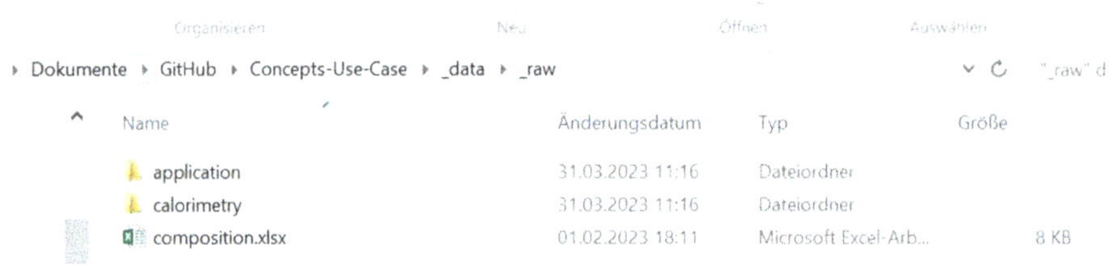
Dokumente › GitHub › Concepts-Use-Case › _data › _raw
Name
Änderungsdatum
Typ
Größe
application
31.03.2023 11:16
Dateiordner
calorimetry
31.03.2023 11:16
Dateiordner
composition.xlsx
01.02.2023 18:11
Microsoft Excel-Arb...
8 KB

Figure 4: Suggested folder structure.

In the given example, our sample identification numbers begin with *P23_X_* followed by a three digit decimal number.[6] This implicitly assumes that no more than 999 experiments will be covered within the scope of this example project. If you are under the impression that this number is not sufficient for your project, feel free to opt for a four- of five- digit number at the end of the identification number.[7]

6 This identification number can easily be combined with a custom built QR code system to avoid typos (see https://pypi.org/project/qrcode/).

7 Unlike the frequently used *coding* approach to filenames, the suggested identification number does not have any inherent meaning. This means that you will not be able to tell from its naming that, e. g., experiment *P23_X_13* has a certain composition. No worries, this information is not lost, we will just get there in a different way.

8 Having the overall concept sketch in mind

Problem

When dealing with the details outlined in the following chapters, it can be challenging, especially for, but not limited to, natural scientists, to loose sight of the "bigger picture".

Context

Typically, R&D scientists are not software developers. And that's fine. Nevertheless, certain skills of the latter are certainly helpful when it comes to dealing with the partially "large[1]" volumes of data obtained from conducting natural science experiments.

Solution

At a very high level, the Data – Information – Knowledge – Wisdom (DIKW)-pyramid[2] as sketched in Figure 8.1 is a helpful model for the scientific process. *Data* is the foundation on the way to *wisdom*. However, it is not exclusively in this first and essential step, where the critical role of R&D scientists shines through. It is only in these stages that they are likely to be most valuable, assuming the *data-driven* approach you are embarking on. Scientist have both the training and the instinct for identifying the appropriate means of characterizing substances and drawing meaningful information from them.

How this extraction of information from experimental data and the building of a larger scale model can be achieved effectively will be the main topic of the following pages. The key idea here is to provide natural scientists with additional tools in the proverbial belt to help them perform repetitive tasks more efficiently and subject to less individual error.

Again, the aim here is to contribute to a more efficient way of *how* scientists can handle their workload, rather than to change the things scientists do. Overall, the techniques presented here are intended to help R&D scientists become even better R&D scientists by leveraging the benefits of some coding to speed up routine processes, thereby freeing up time for more value-added tasks.

Whenever you feel that a chapter herein is "too much", take a step back and find out if the topic covered therein is helpful for the *scientific* topic you are trying to solve. Everything presented here is merely a means

---

1 This is highly dependent on the field in which you work.
2 https://en.wikipedia.org/wiki/DIKW_pyramid

https://doi.org/10.1515/9783111334608-011

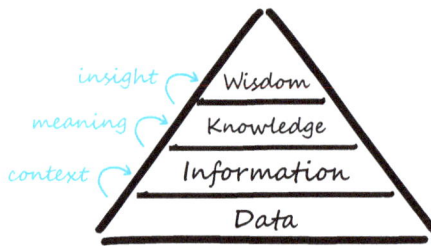

Figure 8.1: Sketch of the DIKW-pyramid. Each level of the pyramid adds value based on initial data, and can be used to answer high-level questions. Importantly, the higher level cannot be reached without a solid foundation from the lower levels.

to an end. If the proposed concept does not support, enable or speed up our process, this chapter may not have been written for the use case you are working on.

9 Initializing a project with poetry

Problem

You want to set up a project for easily sharing with your community. For more details on the sharing aspect, see the outline in Concept 12.

Context

To avoid the common pitfall of "... but it works on my machine" *dependency management* needs to be taken care of properly. In the following, a tool to support this process is presented.

Solution

The `poetry` package is a dependency management and packaging tool for Python. It allows you to declare the libraries your project depends on and it will manage (install/update) them for you. Poetry provides a lockfile to ensure repeatable installs, and can build your project for distribution.[1]

Note that `poetry` should always be installed in a dedicated virtual environment to isolate it from the rest of your system. Under no circumstances should it be installed in the environment of the project that is to be managed by `poetry`. This ensures that `poetry`'s own dependencies are not accidentally updated or uninstalled.

Ideally, a project using `poetry` should be created from scratch. Therefore, open the *Anaconda Prompt* and navigate to the folder, where the package `concepts_use_case` shall be created. To achieve this, run

```
1  poetry new concepts\_use\_case
```

in the command line. The resulting folder structure is shown in Figure 9.1.

---

1 https://python-poetry.org/docs/

https://doi.org/10.1515/9783111334608-012

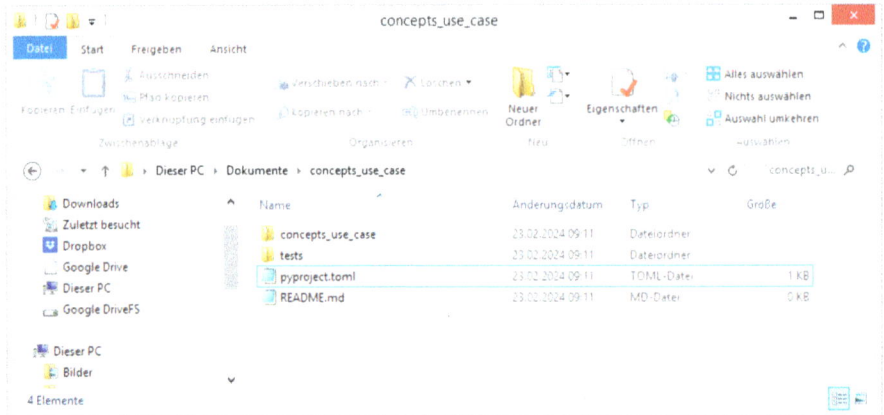

Figure 9.1: Folder structure of the `concepts_use_case` package created with the `poetry new` command.

Discussion

The resulting structure contains
- a *concepts_use_case* folder including an *__init__.py*-file: This is where the source code of the modules to be developed will live. You can also create additional sub-folders corresponding to submodules.
- a *test* folder including an *__init__.py*-file: This is the location suggested by `poetry` for adding tests of the developed modules (see also Concept 15)
- a *README.md* file which is used to share basic information about the purpose of the package, installation guidelines, notes and additional topics worth sharing related to the project, and
- the *pyproject.toml* file: This is file is the most important result of this initialization process. It orchestrates your project and its dependencies. Right now, it looks like this:

To add additional dependencies to the project, use the `poetry add` command with the package you want to use in your project. This will
- modify the *pyproject.toml* file to incorporate the package name and version constraints of the added package, and
- automatically find a suitable version constraint and install the package and its dependencies.

Further reading

For more information on basic use[2] of and managing dependency specifications[3] via poetry, please refer to the online documentation.

---

2 https://python-poetry.org/docs/basic-usage/

3 https://python-poetry.org/docs/dependency-specification/

10 Tracking the environment

Problem

You want to track the virtual environment using version control.

Context

So far, we have created a defined "Python plus packages" combination using conda virtual environments. We have also defined our *Concepts-Use-Case* repo, i. e., the location for storing files relevant for our project. Now it's time to combine these two worlds.

Solution

Open the *Anaconda Prompt* and activate the concept environment via

```
1 conda activate concepts
```

Next, we navigate to the local path of the synchronized repo and export the environment information to a *.yml*-file via

```
1 cd C:\Users\LocalAdmin\Documents\GitHub\Concepts-Use-Case
2 conda env export >concepts.yml
```

The first line changes the directory to the environment path. The second command exports the environment properties to the *concepts.yml* file. In *GitHub Desktop* we see that the file *concepts.yml* has appeared in the list of tracked changes. To make it effective, we follow the commit and push workflows.

Discussion

In the previous steps, we compiled the information on how to build the virtual environment in the *concepts.yml* file and made it subject to version control in the *Concepts-Use-Case* repo. This allows the environment to be restored at a later date and consequently the code to be developed in the future to be re-run – making your current efforts future-proof.

Keep in mind to update the exported *.yml* file regularly after installing additional packages or updating already installed packages.

https://doi.org/10.1515/9783111334608-013

Alternatively, *.yml* files can also be obtained from the *Environments* section of *Anaconda Navigator* using the *Backup* function. Conversely, a virtual environment can be recreated from a given *.yml* file using the *Import* function (see Figure 10.1).

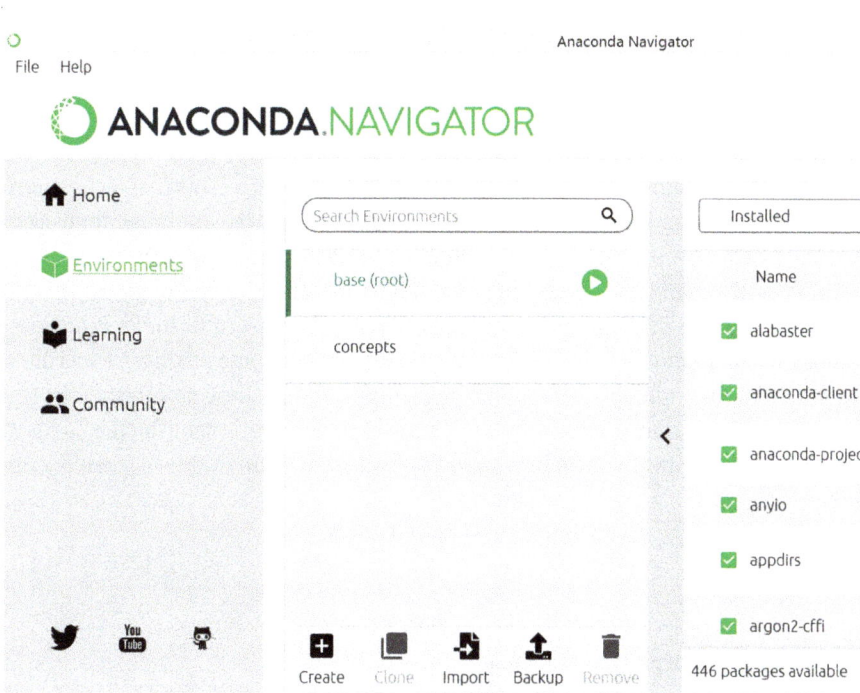

Figure 10.1: Screenshot of the *Anaconda Navigator* showing the options for exporting and importing environment information via *yml*-files using the "Backup" and "Import" functions, respectively.

11 Getting your paths right

Problem

You need to locate files and folders in a particular folder structure.

Context

As mentioned earlier, relying on an organized folder structure can serve as an appropriate way to manage the experimental data generated by a (small- to medium-sized) research project. In a way, it can be thought of as "a poor man's database".

But even if you are fortunate enough to have a database available in your institution, there might be that one case where you will not be able to fit in. Be it a certain type of experiment (not yet) available on the system, or just some auxiliary experiment or side consideration made for certain compositions. Under these circumstances, simply use the information provided by the database as a replacement for the *composition.xlsx* referred to in the introduction to this section as your source of identification numbers.

Solution

Use Python's object oriented file system module `pathlib`.

Listing 11.1: Using `pathlib` for managing directories.

```python
import pathlib

# get current path/working directory
path = pathlib.Path().cwd()

# information
print(f"This_script_is_located_in_path_{path}")

# This script is located in path # C:\Users\LocalAdmin\Documents\
# GitHub\data_playbook_II\_Manuscript\022

# get parent directory
path_parent = path.parent

# define path holding data
path_to_data = path_parent / "_data"

# iterate "_data" contents
for e in path_to_data.iterdir():
    # info
    print(e)

# C:\Users\LocalAdmin\Documents\GitHub\Concepts-Use-Case\_data\_raw
```

https://doi.org/10.1515/9783111334608-014

```
24
25    # check for "nature" of variable
26    print(f"{e}_is_a_directory:_{e.is_dir()}")
27
28    # loop folder contents
29    for i in path_to_data.rglob("*"):
30        # info
31        print(i.name)
32
33    # _raw
34    # application
35    # calorimetry
36    # composition.xlsx
37    # application.json
38    # P23_X_001.xls
39    # P23_X_002.xls
40    # ...
41    # P23_X_020.xls
```

Discussion

The pathlib module offers classes representing file system paths with semantics appropriate for different operating systems.[1]

Navigating one folder level up is as easy as assessing the parent attribute of the path variable. In our case, we want to take a closer look at the contents of the _data folder, which contains our experimental results. To navigate there, the variable path_parent is extended by the separator \ and the string _data.

At the time of writing this piece of code, iterating this path's contents using the iterdir method only returned the folder _raw. The implicitly defined variable e is of type pathlib.WindowsPath. Relying on the is_dir method, confirms that _raw is a directory path.[2]

But that's not what we actually wanted. Rather, we intended to get an overview of all files in the _data folder, i. e., parsing all the possible subfolders. To accomplish this, we use the rglob method, which recursively returns all existing files (of any type, including directories) that match a given relative pattern. Using * as the search pattern will list all file and directory paths. To truncate the actual informative printout, we apply either the name or stem attributes, which return the final path components with and without the last suffix, respectively.

---

1 https://docs.python.org/3/library/pathlib.html

2 Further helpful "is_"-methods are is_file(), is_systemlink() and is_absolute(). For more available methods please refer to the documentation.

Further reading

See the `pathlib`'s documentation for a comprehensive overview of corresponding functions of the `os` and `pathlib` packages.[3]

---

3 https://docs.python.org/3/library/pathlib.html

12 Preparing to share

Problem

You want to be able to easily share the functions and classes you develop as part of a particular project with colleagues or with another project you are working on – not necessarily now, but it's good to plan ahead.

Context

Assuming that you are working in a field of natural scientific R&D where certain tasks are recurring, it is often beneficial to "take the infamous step back", identify common subtasks and break them down into smaller pieces. After some contemplation, you might find that an aspect you are working on in this particular project is relevant also to other parts of your work, or may be relevant to some of your colleagues. As indicated in the problem statement at the beginning of this section, this is not necessarily obvious right from the start of a certain activity. The goal of the present concept, therefore, is to be well prepared in the event that you are struck by this revelation.

Solution

The suggested solution is to use `poetry`[1] for Python packaging and dependency management. It is a highly valuable tool for the entire workflow of developing, building, and publishing Python-packages.

As introduced in Concept 9, a *pyproject.toml* is at the heart of all the `poetry`-associated processes. It contains, among other things, information about the package to be developed, the version number, the author and the required package information, as shown in code section 12.1.

Listing 12.1: pyproject.toml

```
1  [tool.poetry]
2  name = "project23helpers"
3  version = "0.1.0"
4  description = ""
5  authors = ["Matthias_Hofmann_<97462230+mj-hofmann@users.noreply.github.com>"]
6  readme = "README.md"
7
8  [tool.poetry.dependencies]
9  python = "^3.9"
10
```

---

1 https://python-poetry.org/

https://doi.org/10.1515/9783111334608-015

```
11
12  [build-system]
13  requires = ["poetry-core"]
14  build-backend = "poetry.core.masonry.api"
```

This type of file collects all the information poetry requires to build, distribute, and publish your package.

The *pyproject.toml* can be generated
- either using the poetry new command from the console, navigating to the target folder from scratch (see Concept 9),
- or using the poetry init command from the console, navigating to the target folder already holding some files.[2]

Discussion

The previously introduced poetry new and poetry init commands differ in some aspects. The main difference is that it does not create the helpful folder structure, but only a *pyproject.toml* file as specified in Figure 12.1.

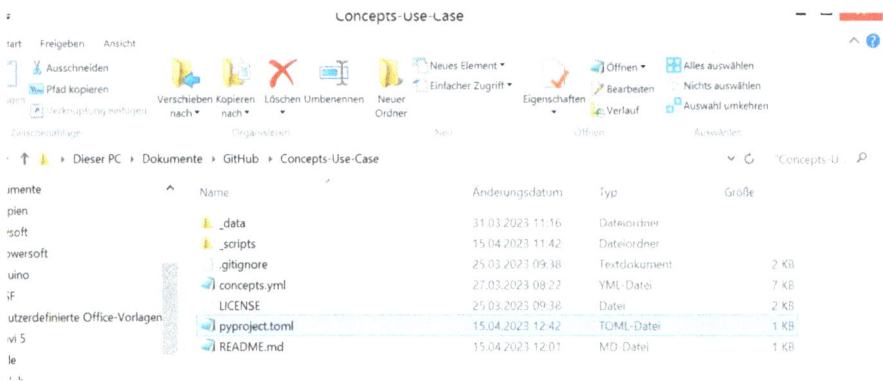

Figure 12.1: Screenshot of the highlighted *pyproject.toml* file within a Python project folder.

In order to mimic the behavior of the poetry new command, just add the directories *tests* and *concept_use_case*[3] each containing an empty *__init__.py* file to yield the folder structure shown in Figure 12.2.

---

2 I used this method creating the *pyproject.toml* file in the already existing and synchronized (see git-concept) *Concepts-Use-Case* folder.

3 The precise naming of the folder is to be taken from the packages keyword given in 12.1.

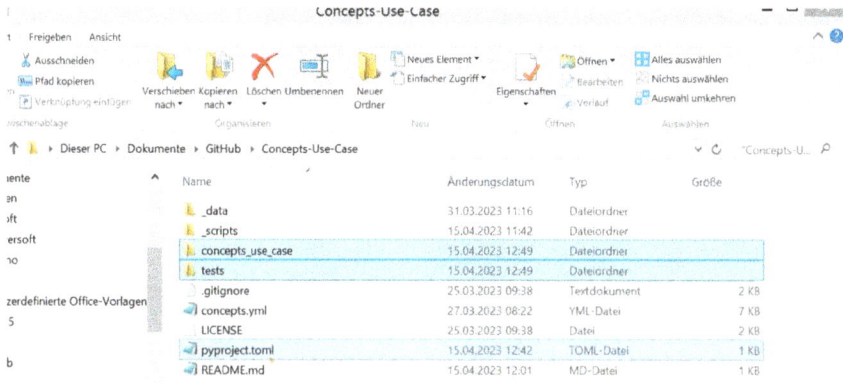

Figure 12.2: Folder structure of the package created with `poetry new`.

Having this structure at hand, organizing your project and considering testing (see Concept 15) will facilitate building your package.

To finally build your package, simply navigate to the folder containing the *pyproject.toml* file and run `poetry build`. This command will create another folder (see Figure 12.3) named *dist* holding both the *.whl* and *.tar.gz* ready to be installed and shared as known from https://pypi.org/.

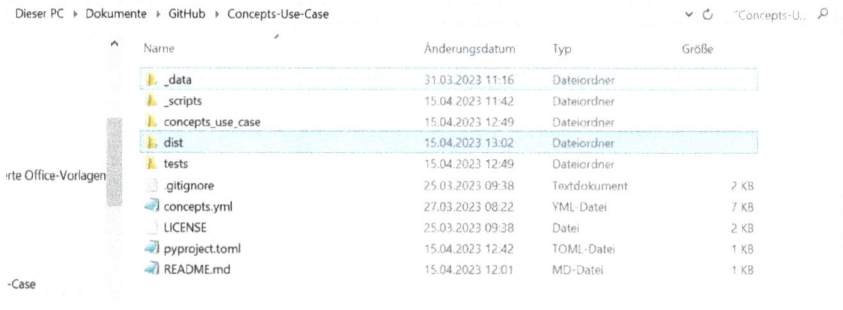

Figure 12.3: Running `poetry build` inside a folder containing a *pyproject.toml* file will create a *dist* folder containing the *source* and *wheel* files.

Further reading

For more information on how to publish packages using `poetry` to either public or private repositories, please refer to the documentation[4] and Concept 45 and Concept 46.

---

4 https://python-poetry.org/docs/repositories/

13 Writing convenience functions

Problem

In your daily work, certain tasks are recurring, such as locating the folder containing experimental data. You don't want to spend time on repetitively writing the same code over and over again.

Context

Repeatedly writing code for basic tasks such as described above in multiple scripts or pieces of code is both tedious and error-prone. It is also a violation of the *Don't repeat yourself (DRY)*-principle of software development.[1]

Solution

Collect this kind of convenience functions in a dedicated module of your package. Typical names might be *utils* or *helpers*. In our example, we want to define a function called `get_data_path` in the `helpers` module that returns the data folder as a `pathlib.WindowsPath`. A possible implementation is shown in code section 13.1.

Listing 13.1: concepts_use_case/helpers.py

```python
import pathlib


def get_data_path(
    use_parent=False, use_parent_parent=False
) -> pathlib.WindowsPath:
    """
    Retrieve path to "_data" based on current working directory

    Returns
    -------
    pathlib.WindowsPath

    """

    # get current path as base assumption
    path = pathlib.Path().cwd()

    if use_parent and not use_parent_parent:
        # use parent of current path
        path = path.parent
```

---

1 https://deviq.com/principles/dont-repeat-yourself

https://doi.org/10.1515/9783111334608-016

```
22    if use_parent and use_parent_parent:
23        # use grandparent of current path
24        path = path.parent.parent
25
26    # append relative subfolder information
27    path = path / "_data"
28
29    # check for type of path
30    if path.is_dir():
31        # info
32        print("\u2713_The_specifed_path_exists_and_is_a_directory.")
33        print(f"___Returning_{path}_as_\"_data\"-path.")
34    else:
35        # raise Exception and show information
36        raise FileNotFoundError(
37            f"The_specified_path_is_not_a_folder_path.\n_data\
38    _____currently_expected_at_{path}."
39        )
40
41    # return path
42    return path
```

Using the function to identify the data path is then as easy as shown in code section 13.2. The path containing the data files is returned as the variable path_data by calling the function get_data_path from the helpers module.

Listing 13.2: Concepts-Use-Case/_scripts/02_writing_convenience_functions.py

```
1   # import custom module
2   from concepts_use_case import helpers
3
4   # get path to data
5   path_data = helpers.get_data_path(use_parent=True)
6
7   #     The specifed path exists and is a directory.
8   #     Returning C:\Users\LocalAdmin\Documents\GitHub\
9   #     Concepts-Use-Case\_data as "_data"-path.
```

Discussion

In the sense of this book, a *convenience function* refers to defining a shorthand version of an action that you expect to do, or that you already perform multiple times across various pieces of code, as a function. Typical examples include, but are not limited to
- checking for the existence of folders and/or files,
- checking the validity of file extensions within a folder,
- validating the hierarchy within a folder against a given assumption,
- validating file names,
- parsing file names to identify patterns,
- ...

The list can be extended. Also note that these functions can easily be converted to methods if you want to opt for an *object oriented* approach in handling measurements corresponding to individual samples.[2]

2 The concept is used in the *tainstcalorimetry*-package for handling heat flow calorimetry data (https://pypi.org/project/tainstcalorimetry/).

14 Using TOML files for configuration

Problem

You want to define a configuration file in a Python-friendly way.

Context

Information about user or project preferences and (user-specific) settings can be handled using configuration files. *Configuration* can, e. g., refer to preferred colors, line widths in scientific plots, font types or sizes, and much more associated with, e. g., a particular user or group of users. To manage the changes required going from one user to another, configuration files are an invaluable tool.

Solution

Build a Tom's Obvious Minimal Language (TOML) file that is "easy to parse into data structures in a wide variety of languages".[1] Interfacing with TOML-files is enabled via the packages tomli[2] or tomllib.[3]

Listing 14.1: Parsing the TOML file *settings.toml*.

```python
1  import tomli
2  import matplotlib.pyplot as plt
3
4  # define filename
5  file = "settings.toml"
6
7  # Opening a Toml file using tomlib
8  with open(file, "rb") as toml:
9      # load
10     toml_dict = tomli.load(toml)
11
12 # define user
13 USER = "UserA"
14
15 name = toml_dict[USER]["name"]
16 user_id = toml_dict[USER]["id"]
17 user_color = toml_dict[USER]["color"]
18 linestyle =  toml_dict[USER]["linestyle"]
19
20 # plot
```

---

1 https://realpython.com/python-toml/
2 https://pypi.org/project/tomli/
3 This module does not support writing TOML.

https://doi.org/10.1515/9783111334608-017

```
21  plt.plot(
22      [0, 0.7, 1], [0, 0.9, 1],
23      color=user_color, linestyle=linestyle
24      )
25  # add title
26  plt.title(f"Plot_for_user_{name}")
```

i Switch to UserB in code section 14.1 and check the corresponding result plot.

Discussion

The parameters, options, settings and preferences applied to operating systems, infrastructure devices and applications as defined in a configuration file can be quite diverse in their nature. Relevant use cases in the world of natural scientific R&D are manifold. To illustrate the use of TOML configuration files, consider the following potential applications.

– You are collecting continuous experimental data for multiple samples using various methods. The samples should be displayed in the same color "across methods". The color mapping can be achieved using a TOML file.

– You are collecting experimental data along with a "reference". The reference data should be displayed in a less conspicuous color and opacity in the sense of the often-cited "guide to the eye". The appearance of "references" can be adjusted using a TOML file.

– You are attempting to log in to a password-protected database or cloud-based storage solution for experimental data using your credentials, e. g., the combination of username and password. To avoid typing the username and password over and over again, they can be stored *locally* within a TOML file and parsed by a piece of code for interfacing with the data source.[4]

Further reading

The `tomli-w` package is the write-only counterpart of `tomli`, providing `dump` and `dumps` functions.[5] As `tomli` is intended to be minimal, it does not include write capability, since most TOML use cases are interpreted as read-only.

---

4 Keep in mind to store this type of file in a private folder that is not accessible to others. Organizations such as industrial research companies and universities typically provide a set of regulations and guidelines on how to leverage this concept.

5 https://pypi.org/project/tomli-w/

15 Getting used to testing

Problem

You want to ensure that functions perform and fulfill actions as intended. This should be true for different inputs specified via variables and keywords.

Context

Sometimes, functions do not behave as initially planned. Reasons for this unexpected behavior include:

- The function is actually incapable of doing what was intended.
- The function is unable to perform the intended actions given the specific input.

Obviously, you want to distinguish between these two scenarios and find a solution as quickly as possible.

Solution

Use the `pytest`-package to test your code. The key idea is to compare (return) values that occur during function execution with your expectations. If a test passes, your code behaves as you expected at least for these specific test cases.

To use the `pytest` package, we need to install it in our example environment *concepts*. Therefore, we open the *Anaconda Prompt*, activate the *concepts* environment, and add the dependency using `poetry` as outlined below.

```
1  conda activate concepts
2  cd C:\Users\LocalAdmin\Documents\GitHub\Concepts-Use-Case
3  poetry add pytest
```

These steps will update the *pyproject.toml*-file. The `pytest`-package is now listed in its dependencies section, as shown in Figure 15.1.

The code section 15.1 below shows a basic example of using `pytest`. To focus on the testing idea, we start by defining the function to be tested, `my_add_function` within the file. In it's simplest application, it may be used to add two integers *a* and *b*. This behavior is tested via the function `test_f`. Taking another look at `my_add_function`, it can also be used to add and subtract float numbers from each other. To ensure this behavior for some more cases, we define the function `test_f_multi` and *parametrize* it to test for multiple cases "in one go".

https://doi.org/10.1515/9783111334608-018

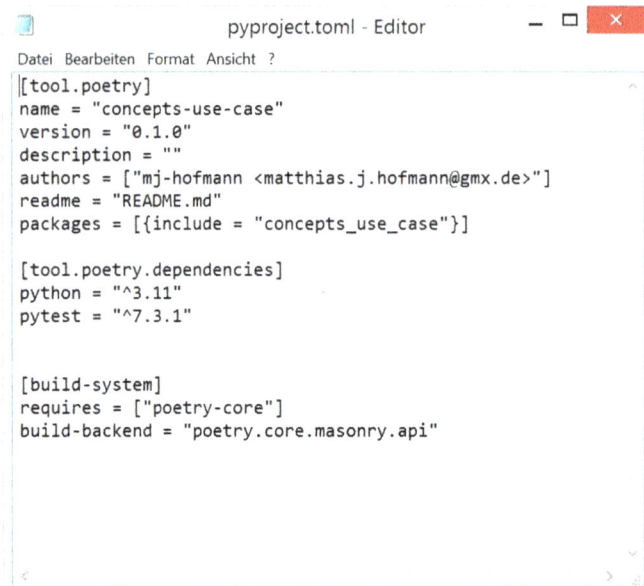

Figure 15.1: Screenshot of the customized TOML file listing `pytest` in the dependencies section.

Listing 15.1: Example showing the basic use of `pytest` via the file `test_code.py`.

```
1  import pytest
2
3  # function to be tested; could also be taken from another module
4  #    via import
5  def my_add_function(a, b):
6      # return sum
7      return a + b
8
9  # test defined function "my_add_function" for one case
10 def test_f():
11
12     # define expected in and outputs
13     a = 1
14     b = 2
15     r = 3
16
17     # use defined function
18     ret = my_add_function(a, b)
19     # check
20     assert ret == r
21
22 # test defined function "my_add_function" for multiple scenarios
23 @pytest.mark.parametrize(
24     "a, b, result", [
25         (1.2, 2.6, 3.8),
26         (3.7, -1.1, 2.6),
27         (2, 4, 6),
28         (-1.1, -0.9, -2),
```

```
29          (-1.1, -0.9, -2.0),
30  ])
31  def test_f_multi(a, b, result):
32      # use defined function
33      ret = my_add_function(a, b)
34      print(ret)
35      # check
36      assert ret == result
```

To finally run the test, we navigate to the folder of the file test_code.py and run the pytest command in the console (see Figure 15.2).

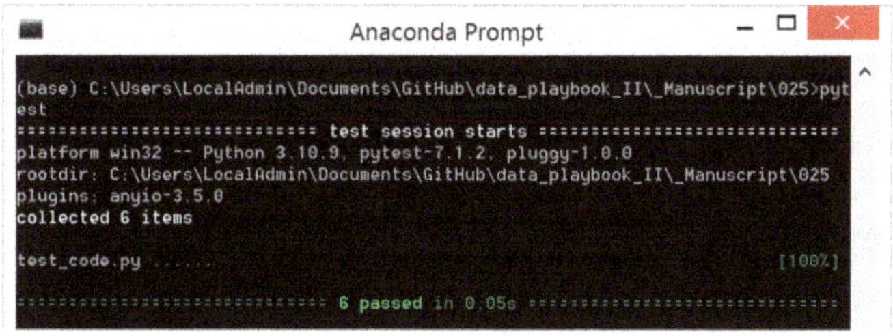

Figure 15.2: Screenshot of the console after executing the pytest command. Success is confirmed by the "6 passed" message.

Modify the my_add_function to return an integer value and rerun pytest.

Discussion

The pytest framework makes it easy to write small, readable tests, and can scale to support complex functional testing for applications and libraries.[1] Certainly, this package has a lot more to offer than shown in the few lines above.

1 https://docs.pytest.org/en/7.4.x/

> **i** Running `pytest` without mentioning a file name will run all files of the format *test_*.py* or *_test.py* in the current directory and subdirectories.

While `pytest` is mostly used for API testing, it's application is also beneficial in the realm of natural sciences. Potential use cases include testing the parsing of experimental result files from different generations of instruments or control software versions. Additionally, exported files can be modified and adapted by the respective users. In order to ensure a proper functioning of your envisioned functions, `pytest` can be an essential tool for you.

Further reading

More detailed introductions to testing with `pytest` are available in online tutorials[2] and in the project's documentation.[3]

---

2 https://www.tutorialspoint.com/pytest/pytest_introduction.htm
3 https://docs.pytest.org/en/7.4.x/

Interfacing with common data formats

Experimental data comes in many forms. Depending on the setup and devices used, the results may be in the form of *xlsx, csv, txt* or others.

The following chapters present approaches for interfacing with files of each type.

In all cases, the key idea is to treat the respective file as a data source, extract the parts that are considered relevant, and compress them into a table.

https://doi.org/10.1515/9783111334608-019

16 Reading Excel files

Problem

You want to parse the contents of an Excel file.

Context

Numerous experimental devices provide the ability to export results to an Excel-type format. This is true for analysis methods across the analytical spectrum from thermal analysis to physical characterization to spectroscopy and wet chemical methods.

In addition to the wealth of scientific methods, there are numerous sub-types of Excel files such as the older *xls* files. Irrespective of their experimental source and precise place in the evolution of Excel-type files, they can be considered tabular if a (more or less large) part of the file content is discarded.

Solution

Use the `read_excel` function from pandas to parse Excel files. By specifying the appropriate keywords, tabular data can be extracted from the respective source files in the *intended*, i. e., tabular form.

In the simplest case, data can be parsed from an Excel file using a functional approach, i. e., by calling the `read_excel` function and specifying the appropriate keywords. Therefore, it is recommended to open the file in Microsoft® Excel® or any other spreadsheet tool and take a look at the relevant parts (see Figure 16.1). The extraction of the contents of the raw data sheet in tabular form is demonstrated in code section 16.1.

Figure 16.1: Screenshot of the opened Microsoft® Excel® file to identify the parts considered relevant.

https://doi.org/10.1515/9783111334608-020

Listing 16.1: Reading the contents of an Excel file using the functional approach.

```
1  import pandas as pd
2
3  # define file
4  file = "sample_data.xlsx"
5
6  # read from sheet "Raw data"
7  raw_data = pd.read_excel(
8      io=file,
9      sheet_name="Raw_data",
10     header=[0,1]
11     )
12
13 # "flatten" multi-index column names
14 raw_data.columns = ["_".join(i) for i in raw_data.columns]
```

In addition, an object-oriented approach can be used to parse Excel files via pandas. This will be demonstrated in the following. The keywords of the call in the parse-method match the keywords used in the above case. Again, the data can be transferred to a tabular pd.DataFrame as shown in code section 16.2.

Listing 16.2: Reading the contents of an Excel file using the object oriented approach.

```
1  # initialize object
2  xl_file = pd.ExcelFile(file)
3
4  # show available sheets
5  for sheet in xl_file.sheet_names:
6      # info
7      print(sheet)
8
9  # Experiment info
10 # Raw data
11
12 # parse "Experiment info" sheet
13 info = xl_file.parse(
14     sheet_name="Experiment_info",
15     header=None
16     )
```

Discussion

Typical sources of errors and/or unexpected behavior include the use of the following keywords in the pd.read_excel function:
- *decimal* defines the character to use as the decimal separator. Use the comma for European style data.
- *engine* is used to interface with the various Excel-type formats available, such as *xlsx, xls ,odf, ods, odt.*

– *header* defines the zero-indexed row number to use as the column header of the resulting `pd.DataFrame`.

Further reading

The full documentation of the `read_excel` function is available on https://pandas. pydata.org/docs/reference/api/pandas.read_excel.html.

17 Reading text files

Problem

You want to parse the contents of a text file.

Context

Numerous experimental devices provide the ability to export results to a text-type format. This is true for analysis methods across the analytical spectrum from thermal analysis to physical characterizations to spectroscopy and wet chemical methods.

In addition to the wealth of methods, there are plenty of sub-types of text files such as *csv* files. Further complexity is introduced by machine vendors creating other file types with custom extensions. To check if you are indeed dealing with a text-type file, try opening the target file with a text editor of your choice.[1]

Unlike the previously discussed Excel-type result files (see Concept 16) obtained from experimental devices, text files do not necessarily come in a tabular format that can be easily converted to the desired pd.DataFrame structure.

Solution

There are at least two ways to parse the contents of a text file, as shown below.

The first one relies on using the read_csv-function of the pandas package. By specifying the appropriate keywords, tabular data can be extracted from the respective source files in the *intended* form. This requires a tabular data block at a known location within the source file. In the example shown in Figure 17.1 and code section 17.1, we assume that this data block starts at line 17.

Change the line number declared as header to values other than 16 and check the consequences.

Listing 17.1: Reading the contents of a text file using the functional approach.

```
1  import pandas as pd
2
3  # define file
4  file = "sample_data.csv"
5
6  # read via pandas function
```

---

1 Notepad ++ (https://notepad.plus/) is a recommended solution that comes with a lot of functionality.

https://doi.org/10.1515/9783111334608-021

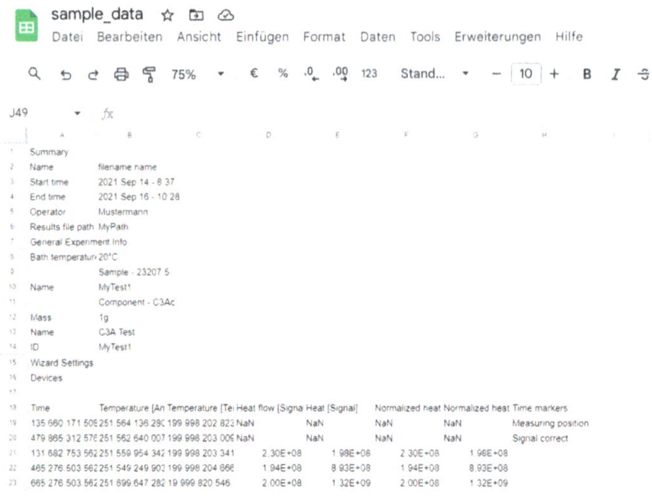

Figure 17.1: Screenshot of the opened *csv* file to identify the parts that are considered relevant.

```
7   data_1 = pd.read_csv(
8       filepath_or_buffer=file,
9       sep=",",   # define column separator
10      header=16 # define row to be used as header
11      )
```

The second approach to parsing text data relies on initially reading the file contents as strings and then doing some subsequent processing on these readings. The contents variable created here is a list of strings. In order to identify the "data block" in this approach, we need to specify the column separator (which is the same as in the previous method) and set our expectation for the number of columns within a data row. Here, we want a data row to consist of precisely eight columns. The last lines of code section 17.2 demonstrate how the list of data rows from the original text file is converted into a pd.DataFrame.

Listing 17.2: Reading the contents of a text file as string.

```
1   with open(file, mode="r") as f:
2       # read
3       contents = f.readlines()
4
5   # define column separator
6   col_sep = ","
7
8   # restrict list of strings to "data lines"
9   contents = [i for i in contents if len(i.split(col_sep)) == 8]
10
11  # convert list of strings to DataFrame
12  data_2 = pd.DataFrame(
13      data=[i.split(col_sep) for i in contents[1:]],
```

```
14        columns=[i.strip("\"\n") for i in contents[0].split(col_sep)],
15        )
16
17   # type conversion
18   for c in data_2.columns:
19        # try conversion to float
20        try:
21             data_2[c] = data_2[c].astype(float)
22        except:
23             pass
24
25   #              Time    ...         Time markers
26   # 0         0.013566   ...    "Measuring position
27   # 1         0.479865   ...       "Signal correct
28   # 2         1.316828   ...
29   # 3         4.652765   ...
30   # 4         6.652765   ...
31   #              ...     ...
32   # 1112    5931.300226  ...
33   # 1113    5941.498468  ...
```

Discussion

Typical sources of errors and/or unexpected behavior include the use of the following keywords in the read_csv function:
–   *decimal* defines the character to be used as decimal separator. Use the comma for European style data.
–   *encoding* defines the encoding to use when reading or writing.[2]

Further reading

The full documentation for the read_csv function is available on https://pandas.pydata.org/docs/reference/api/pandas.read_csv.html.

---

2 The list of Python standard encodings is summarized at https://docs.python.org/3/library/codecs.html#standard-encodings

18 Reading text from Word files

Problem

You want to parse the contents of a Word file.

Context

Along with Microsoft® Excel® and Microsoft® Power Point®, Microsoft® Word® remains a frequently used tool for the documentation of experiments. Thanks to its ease of use and its many formatting and highlighting capabilities, it can – under certain circumstances – be an appropriate choice for documenting experimental processes in text form.

Historically, it was common practice to collect numerous protocols of exactly the same format to document the compositions and behaviors within, e. g., concentration series. For ease of use, documentation templates in a Word format can be applied to support this type of workflow. In this and the following concept (see Concept 19), the parsing of the contents of individual Word files is demonstrated.

In addition, the contents of these file types can be formatted as (among other things) text or tables.[1] Each of these can be parsed and compiled for further analysis.

Solution

To parse the textual content of Word files, the *python-docx*[2] package is a good starting point.

A screenshot of the Word file to be parsed in given in Figure 18.1. On a high level, it contains some text outside the table and some more text inside the cells of the table.

Here we want to extract the *text* part of the Word document. As shown in code section 18.1, a new docx.document.Document object is instantiated based on the file path. The content is accessible through the text attribute of the paragraphs. In order to obtain the content of the file as a single string, the text of the paragraphs can be combined as demonstrated in code section 18.1.

Listing 18.1: Reading the contents of the Microsoft® Word® file *sample.docx* using the python-docx-package.

```
1 # https://pypi.org/project/python-docx/
2 from docx import Document
```

---

1 The discussion of accessible content types is limited here to these two cases.

2 https://pypi.org/project/python-docx/

https://doi.org/10.1515/9783111334608-022

Hi there,

I'm a test document containing a table

column A	column B
cell 1	cell 2
cell 3	
	cell 4

... and some text.

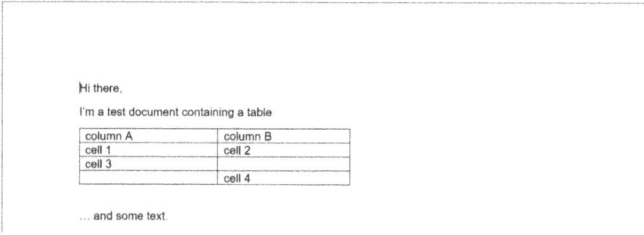

Figure 18.1: Screenshot of the Microsoft® Word® file containing text and tabular data as an example of an experimental protocol.

```
3
4  # define file
5  file = "sample.docx"
6
7  # init
8  doc = Document(file)
9
10 # init list of paragraphs
11 list_p = []
12
13 # get text content
14 for paragraph in doc.paragraphs:
15     # append text to list
16     list_p.append(paragraph.text)
17
18 # compile contents as one string
19 content_text = "\n".join(list_p)
20
21 # show text content
22 print(content_text)
23
24 # Hi there,
25 # I'm a test document containing a table
26
27 # ... and some text.
```

Discussion

While Microsoft® Word® may not seem like the most obvious choice for documenting experiments, it comes along with some positive aspects to take into account. Assuming that you are required – for whatever reason – to provide a highly polished documentation of an experimental process, including specific formatting requirements, Microsoft® Word® might actually be a *natural* choice for this purpose, since it is available on most notebooks and used throughout research facilities "at no additional cost" to the user and is intuitive to use. It allows for formatting and a vast amount of flexibility to account for the specifics of a particular experiment. Also, the separation between textual and tabular data can be clearly drawn. Furthermore, the ability to easily create templates to be filled in later is appealing to many.

The trouble begins when the format of a template changes during the course of an experiment. Assuming that a set of experiments will not be too exhaustive[3] and will not take too much time, so that you are confident that you can work with *this* template, Word files can be a perfectly fine type of documentation.

Additionally, the recent availability of Artificial Intelligence (AI) can even contribute to the role of Microsoft® Word®-like files for documenting experimental results in a textual form. The collected text in combination can, e. g., be submitted to an AI tool to summarize or transform the content in a certain way.

Further reading

An alternative method for collecting the contents of a Microsoft® Word® file is based on the packages `zipfile` and `ElementTree`.[4] The use of the packages `docx2txt`, `docx`, and `docx2python` is also recommended.[5]

---

3 for whatever that means.

4 https://dadoverflow.com/2022/01/30/parsing-word-documents-with-python/

5 https://theautomatic.net/2019/10/14/how-to-read-word-documents-with-python/

19 Reading tables from Word files

Problem

You want to parse the tabular contents of a Word file.

Context

As mentioned in the previous concept (see Concept 18), Microsoft® Word® files can be used to document experimental procedures and results in great deatil and rich formatting.

Besides comments and descriptive text, tabular data is an integral part of scientific data handling. Examples include tables describing the composition of formulations and collections of characteristic parameters corresponding to those formulations. In the following, the extraction of tabular data into a pd.DataFrame from the sample Microsoft® Word® *sample.docx* file will be demonstrated.

Solution

As in Concept 19, we rely on the python-docx[1] package to parse the contents of a Word file.

Here we want to extract the *tabular* part of the Word document. As shown in code section 19.1, a new docx.document.Document object is instantiated based on the path of the Microsoft® Word® file. From this, docx.table.Table objects can be drawn one by one. In the present example file, we will focus on the first table. Finally, the contents of the table cells are combined into a pd.DataFrame (see Figure 19.1), the target structure for many of the considerations throughout this book.

Listing 19.1: Reading the tabular contents of the Microsoft® Word® file *sample.docx* using the python-docx-package.

```
1  # https://pypi.org/project/python-docx/
2  from docx import Document
3  import pandas as pd
4
5  # define file
6  file = "sample.docx"
7
8  # init
9  doc = Document(file)
10
11 # init list of table contents as tuple
12 list_t = []
```

---

1 https://pypi.org/project/python-docx/

https://doi.org/10.1515/9783111334608-023

```
13
14  # use first table within the file
15  table = doc.tables[0]
16
17  # get table contents
18  for row in table.rows:
19      # get the row's cells as text via list comprehension
20      content_row = [cell.text for cell in row.cells]
21      # append the tuple to list
22      list_t.append(content_row)
23
24  # convert list to pd.DataFrame
25  content_table = pd.DataFrame(
26          list_t[1:],  # use contents from 2nd row on as table body
27          columns=list_t[0]  # use first row as header
28          )
29
30  # show text content
31  print(content_table)
32
33  #     column A column B
34  # 0    cell 1   cell 2
35  # 1    cell 3
36  # 2             cell 4
```

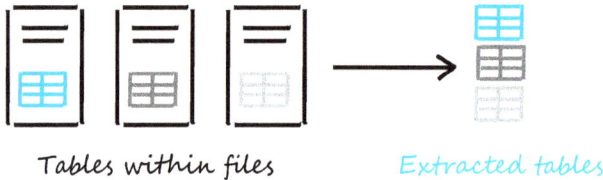

Tables within files   Extracted tables

Figure 19.1: Conceptual extraction and combination of table contents across multiple Microsoft® Word® source files to create an "overall" table.

Discussion

To ensure the extraction of the same type of tables across multiple Microsoft® Word® files, they need to be built according to the same principles, i. e., having the same number and names of columns to keep things simple. For example, the first table in a Word document could contain information about the composition and the second table in a Word document might contain information about particle size characteristics. To allow for easy concatenation of tables collected across multiple files, the tables should have the same number of columns and, most importantly, the same column names.[2]

---

2 Take care with upper and lower case characters as well as with trailing white spaces. Any aspect that is not taken into account properly in the individual source files will require additional effort in the subsequent steps.

20 Reading PDF files

Problem

You want to parse the contents of a PDF file.

Context

There's a variety of experimental devices that have the ability to create reports in PDF format. This practice ifs often encountered in the field of service analytics, as carried out by central analytics departments of larger research facilities.

In this scenario, (internal) clients or customers provide one or multiple samples to an analytics team requesting to perform a characterization according to a specific Standard Operating Procedure (SOP). From the client's perspective everything[1] is known, except the experimental values corresponding to the individual samples.

The report resulting from this analysis request may be in the form of a PDF for documentation purposes[2] summarizing information on instrument settings, tables of experimental raw data, and characteristics derived from the latter.

As these types of reports are regularly generated by automated processes or by software that comes with the experimental equipment, they can be parsed – at the receiving end – to extract *data* and *information* (see Concept 22).

Solution

The PyPDF2-package[3] provides a convenient way for reading the textual content of a PDF file into a str variable.

As shown in code section 20.1, a PyPDF2._reader.PdfReader object is generated by specifying the path to the PDF. Using a simple for-loop, the textual content of each page is extracted and appended to the text variable of type string.

Listing 20.1: Reading the contents of a PDF file using the PyPDF2 package.

```
1  # importing required modules
2  import PyPDF2
3
```

---

1 In terms of the experimental procedure including sample preparation, device settings, etc.

2 Requirements can arise from organizational policies, agreements between companies, or the need to meet certification requirements demanded by other regulatory bodies. In some cases, organizations are required by law to provide dated proofs of analysis.

3 https://pypi.org/project/PyPDF2/

https://doi.org/10.1515/9783111334608-024

```
 4  file = "sample1.pdf"
 5
 6  # %% read
 7
 8  # creating a pdf file object
 9  pdfFileObj = open(file, 'rb')
10
11  # creating a pdf reader object
12  pdfReader = PyPDF2.PdfReader(pdfFileObj)
13
14  text = ""
15
16  # loop pages
17  for page_num, page in enumerate(pdfReader.pages):
18      # info
19      print(f"Reading text from page {page_num+1}.")
20
21      # extracting text from page
22      text += page.extract_text()
```

Concept 22 demonstrates how to extract the desired pd.DataFrame structure from this *str*-type variable using regular expressions.

Discussion

There are certainly a number of packages available on The Python Package Index (PyPI) that are capable of parsing PDF-files. For the process of parsing the textual content of a PDF and transforming this content considered relevant into a pd.DataFrame for further processing described here, combining the packages PyPDF2 and re has proven to be reliable and sufficiently powerful.

Further reading

PyPDF2 is a free and open source pure Python PDF library capable of splitting, merging, cropping, and transforming the pages of PDF files. It can also add custom data, viewing options, and passwords to PDF files. PyPDF2 can also retrieve text and metadata from PDFs.[4]

Alternative packages for working with PDF files include PyMuPDF[5] and pikepdf.[6]

---

4 https://pypi.org/project/PyPDF2/
5 https://pypi.org/project/PyMuPDF/
6 https://pypi.org/project/pikepdf/

21 Parsing website contents

Problem

You want to parse the (tabular) content of a website.

Context

There are many lists on the Internet that are relevant to scientists in research and development. Many of them are related to regulatory issues.

One example is the so-called *Candidate List of Substances of Very High Concern for Authorisation.*[1] This list published by the European Chemicals Agency (ECHA), is updated regularly and can have serious consequences for the direction of a scientist's research due to the legal implications associated with it.

In short, if a substance you are working with in the lab appears on this list, you will no longer be allowed to use it in commercial products above a certain threshold concentration. In an industrial research setting, this comes along with requirements to inform customers, partners and – most importantly for us scientists – future focus areas of research. So you definitely want to know, that the list has been extended.

Solution

Use the read_html-method of pandas to parse the tabular content of websites. Given the url, the method returns a list of pd.DataFrames contained on the website. According to code section 21.1, we select the first DataFrame on this page as our content of interest.

Listing 21.1: Reading tabular contents of a html website.

```
import pandas as pd

# define url
url = "https://echa.europa.eu/en/candidate-list-table"

# read
echa_candidates = pd.read_html(url)[0]

# url of "second page"
fragment_url = "https://echa.europa.eu/en/candidate-list-table?p_p_id\
=disslists_WAR_disslistsportlet&p_p_lifecycle=0&p_p_state=normal&p_p_\
mode=view&_disslists_WAR_disslistsportlet_haz_detailed_concern=&_\
disslists_WAR_disslistsportlet_orderByCol=dte_inclusion&_disslists_\
WAR_disslistsportlet_substance_identifier_field_key=&_disslists_WAR_\
```

---

1 https://echa.europa.eu/de/candidate-list-table

https://doi.org/10.1515/9783111334608-025

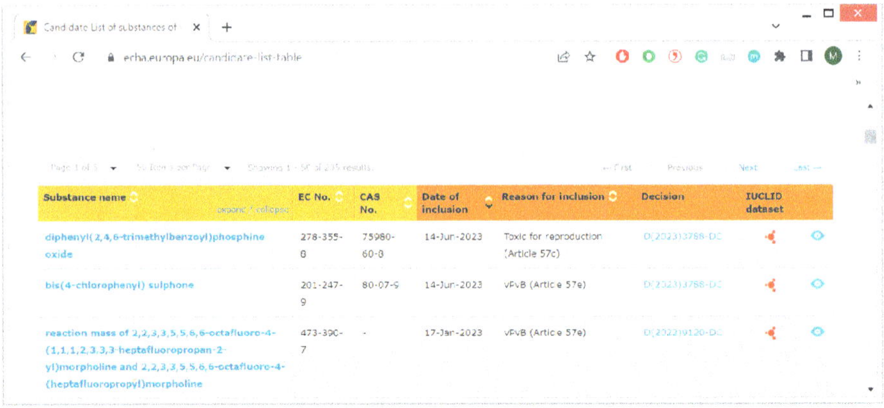

Figure 21.1: Screenshot of the ECHA Candidate List on the official website.

Checking the resulting pd.DataFrame we see, that is does not contain the full set of 235 entries referenced in the table header (see Figure 21.1).

Why is this the case? Apparently, only 50 items are displayed on one page. To see the "next page", we have to press the "Next" button at the bottom right. On clicking, we see a redirect to a much longer Uniform Resource Locator (URL). At a conceptual level, multiple parameters are appended to the base URL to redirect to the other parts of the table. By navigating through the pages, we see that the number after the "cur" parameter in the URL changes. In order to obtain the complete list of all chemicals, we collect the DataFrames corresponding to the individual URLs and finally combine them using the pd.concat method as shown in code section 21.2.

Listing 21.2: Combining tabular contents of pd.DataFrames obtained from multiple html websites.

```
1  # url of "second page"
2  fragment_url = "https://echa.europa.eu/en/candidate-list-table?p_p_id\
3  =disslists_WAR_disslistsportlet&p_p_lifecycle=0&p_p_state=normal&p_p_\
4  mode=view&_disslists_WAR_disslistsportlet_haz_detailed_concern=&_\
5  disslists_WAR_disslistsportlet_orderByCol=dte_inclusion&_disslists_\
6  WAR_disslistsportlet_substance_identifier_field_key=&_disslists_WAR_\
7  disslistsportlet_delta=50&_disslists_WAR_disslistsportlet_orderByType\
8  =desc&_disslists_WAR_disslistsportlet_dte_inclusionFrom=&_disslists_\
9  WAR_disslistsportlet_dte_inclusionTo=&_disslists_WAR_disslistsportlet\
10 _doSearch=&_disslists_WAR_disslistsportlet_deltaParamValue=50&_\
11 disslists_WAR_disslistsportlet_resetCur=false&_disslists_WAR_disslist\
12 sportlet_cur="
13
14 # init list of candidates
15 list_candidates = []
16
17 # loop
18 for i in range(1,10):
19     try:
20         # get
21         candidates = pd.read_html(f"{fragment_url}{i}")[0]
```

```
22      except IndexError:
23          # break loop
24          break
25
26      # add to list
27      list_candidates.append(candidates)
28
29  # %% build overall dataframe
30  candidates = pd.concat(list_candidates)
31
32  # discard not-NaN "Substance name" rows
33  candidates = candidates.dropna(subset=["Substance_name"])
```

Discussion

Web scraping, i. e., the collection of data available on the internet for specific purposes is a field of its own. In addition to technical aspects, there are also legal considerations to be made, especially if there's an intention to use personal data for commercial purposes. This clearly should and cannot be main topic here.

As outlined above, some basic web scraping should be in the toolbox of a modern researcher to automate some – to be honest – tedious work such as looking up lists on various webpages and free up additional time for more creative and value-adding work.

Further reading

A detailed introduction to approaches to *true* web scraping is given in *Web Scraping with Python*.[2]

On https://realpython.com/beautiful-soup-web-scraper-python/ a basic introduction to the bs4-package is provided. The latter aims to make it easy to scrape information from web pages.[3]

---

2 https://www.oreilly.com/library/view/web-scraping-with/9781491985564/
3 https://pypi.org/project/beautifulsoup4/

22 Leveraging regular expressions

Problem

You want to extract *relevant*[1] parts of unstructured text data.

Context

Referring back to Concept 20, where the textual data of the PDF-file was collected as a `str`-type variable, we set ourselves the task of extracting the table of experimental results as a pd.DataFrame. The latter can serve as the basis for further plotting and analysis steps.

Looking at the text data extracted from the PDF (see Figure 22.1), it is visually easy to identify the *data part*.

```
text_example.txt        ×   +
Datei   Bearbeiten   Ansicht
[Massenkonzentration]) -27,543038 [Δσ/Δ lg(c)]Links Mittlere Probentempera
ΓPlatzbedarf eines Moleküls an der Oberfläche nm² 0,34444660 AGibbs freie
Mizellkonzentration mN/m 28,661 σCMCKritische Mizellkonzentration mmol/L -
NameZusammenfassungMessdaten 30,0 mL Volumen 0,997 g/cm³ Dichte 1,0 g/cm³
Vorlagelösung -- Molare Masse 55,4089 mol/L Molarität Vinapor Name water N
Mittl. Temperatur
[°C]Mittl. GFS [mN/m] Konzentration
[mg/L]Abgelaufene Zeit [s] Zyklusnummer
25,23 23,86 5000,0000 84,64 1
25,27 23,75 4721,0034 161,92 2
25,26 23,90 4457,5747 249,19 3
25,29 24,03 4208,8451 325,25 4
25,25 24,12 3973,9944 400,83 5
25,25 24,14 3752,2482 488,27 6
25,24 24,28 3542,8753 563,16 7
25,24 24,31 3345,1853 651,28 8
25,28 24,44 3158,5263 728,99 9
25,27 24,51 2982,2827 803,91 10
25,27 24,50 2815,8733 892,16 11
25,24 24,76 2658,7495 969,12 12
25,26 25,22 2510,3931 1059,52 13
25,28 25,70 2370,3149 1137,96 14
25,26 26,04 2238,0530 1213,50 15
25,27 26,34 2113,1711 1301,46 16
25,26 26,85 1995,2576 1377,21 17
25,25 27,26 1883,9236 1466,22 18
25,30 27,69 1778,8020 1541,51 19
25,30 28,13 1679,5461 1615,58 20
25,29 28,53 1585,8285 1701,68 21
25,27 29,23 1497,3404 1778,40 22
25,29 29,71 1413,7898 1865,70 23
```

Figure 22.1: The `str` variable to which regular expressions apply, as seen through a text editor.

[1] Necessarily, the term *relevant* is highly context and use-case dependent. To get an idea, a relevant part of unstructured data might be a section *read as* a table of experimental raw data.

https://doi.org/10.1515/9783111334608-026

Solution

The re-package makes it easy to identify those visually apparent "data lines". The general sequential steps are as follows:

- Definition of a *pattern* to be searched for in the original string. This pattern defines how we expect a data line to appear in the text.
- Looking for this pattern in the string and collecting the returned values as a list.
- Building a pd.DataFrame based on the previously obtained list of values and pattern.[2]

Listing 22.1: Leveraging regular expressions via the re package to identify experimental data values from a str-type variable and their transformation to a pd.DataFrame.

```python
1  import re
2  import pandas as pd
3
4  # define pattern using named capturing groups
5  pattern = re.compile(
6      "^(?P<T_C>\d\S+)\s(?P<gamma_mN_m>\S+)\s(?P<c_mg_L>\S+)\
7  \s(?P<t_s>\S+)\s(?P<P5>\d+)$",
8      flags=re.M
9      )
10
11  # search through
12  findings = re.findall(pattern, text)
13
14  # list of tuples to DataFrame
15  values = pd.DataFrame(
16          findings,
17          columns=pattern.groupindex.keys()
18          )
19
20  # convert
21  for _c in values.columns:
22      values[_c] = values[_c].str.replace(",", ".").astype(float)
```

Discussion

Regular expressions are certainly not the most intuitive way to extract information from text, but they are an extremely powerful tool. Compared to the typical use cases of regular expressions for validating postal codes, telephone numbers or email addresses, regular expressions for applications in natural scientific R&D are typically rather easy to write, as the source is usually highly structured – probably in a rather unexpected way.

---

2 In the example shown, *named capturing groups* are used to avoid the complications of index-based mapping column values and column names.

In short, if is easy for you as a human reader to distinguish *data lines* from *information lines*, it should be fairly straightforward to translate this into one or multiple regular expression patterns.

Further reading

Of course, there are considerations for using generative transformational models to build regular expressions.[3]

As natural scientists, however, typically like to understand how things are done and work, investing some time in understanding the basics of building regular expressions is far from a waste of time. For a more detailed description of how to build regular expressions for parsing natural science experimental result files see *Data Management for Natural Scientists*.[4,5]

3 https://blog.enterprisedna.co/chatgpt-for-regular-expressions/

4 https://www.degruyter.com/document/doi/10.1515/9783110788433/html?lang=en

5 https://github.com/mj-hofmann/Data-Management-for-Natural-Scientists

23 Writing to a database

Problem

You want to collect data and information associated with a particular *object* in a database. This can be considered the reverse way of the process described in Concept 24.

Context

In a typical setting, the above mentioned "object" can be a formulation or mixture of compounds originating either from a chemical reaction, a physical mixture or a combination of both. *Information* related to the latter could be, e. g., insights into when it was created, by whom it was created, the project in connection with which is was created, or information on how much of the material has been created. Associated *data* can come in the form of experimental result files generated during the characterization using the numerous methods available. Here, *data* refers to the unprocessed files as obtained from the experimental equipment without further processing or user interference. Depending on the manufacturer and the type of analysis, these results may come in any file format, from text files (csv, txt) to binary files (pdf, xlsx, proprietary formats of individual machine vendors).

Solution

To allow for a "sorting in" of data and information corresponding to an individual experiment, raw material, or mixture of raw materials, the creation of a database and the table it contains is demonstrated in code section 23.1.

A local SQLite-database[1] named `example.db` is created to introduce the concept. Within this database, the *table* named *experiments* is defined. For display purposes, this table contains four columns:

- The *id* column serves as the table's primary key, i. e., it is unique within the table.
- The *file_pdf_name* stores text-type values and serves to collect the names of the Binary large object (BLOB)-type data stored in the *file_pdf* column.
- The *file_pdf* stores BLOB-type data such as unprocessed experimental raw data files obtained from analysis devices or images.
- The *experiment_name* stores a human-readable name for the experiment.

---

1 The use of SQLite-databases comes naturally with Python. Irrespective of the choice of the database, the concepts outlined here can be applied to other types of databases.

https://doi.org/10.1515/9783111334608-027

Listing 23.1: Creating a SQLite-database and a table within it for storing data and information corresponding to an object, e. g., a formulation or reaction composition.

```
1  import sqlite3
2
3  try:
4      sqlite_connection = sqlite3.connect("example.db")
5      cursor = sqlite_connection.cursor()
6
7      # create table
8      statement_create_table = '''CREATE TABLE experiments (
9                                  id INTEGER PRIMARY KEY,
10                                 file_pdf_name TEXT NOT NULL,
11                                 file_pdf BLOB NOT NULL,
12                                 experiment_name TEXT NOT NULL
13                                 );'''
14
15     # execute table creation
16     cursor.execute(statement_create_table)
17     sqlite_connection.commit()
18     print("SQLite table created")
19
20
21 except sqlite3.Error as error:
22     print("Error while connecting to sqlite:", error)
23
24 finally:
25     if sqlite_connection:
26         sqlite_connection.close()
27         print("The SQLite connection is closed")
```

Once created, we need to populate the `experiments` table with data and information related to our example experiments as described in code section 23.2. For convenience, we define an `insert`-function that takes the primary key, the name of the experiment, and the file to insert as arguments. The PDF files used in this example can certainly be exchanged for images, Microsoft® Excel® files, and other file types in real-world use cases.

Listing 23.2: Writing data and information to the SQLite-database *example.db*.

```
1  import sqlite3
2  from pathlib import Path
3
4  # helper function to convert binary data
5  def convert_to_binary_data(filename):
6      # Convert to binary format
7      with open(filename, "rb") as file:
8          # extract data
9          blob_data = file.read()
10     # return
11     return blob_data
12
13
14 # reconsider target table's structure
15 # CREATE TABLE experiments (
```

```
16  #     id INTEGER PRIMARY KEY,
17  #     file_pdf_name TEXT NOT NULL,
18  #     file_pdf BLOB NOT NULL,
19  #     experiment_name TEXT NOT NULL
20  #     );
21
22  def insert(experiment_id : int, experiment_name : str,
23              pdf_name : Path):
24      try:
25          sqlite_connection = sqlite3.connect("example.db")
26          cursor = sqlite_connection.cursor()
27          print("Connected_to_SQLite")
28
29          # define query
30          sqlite_insert_blob_query = """INSERT INTO experiments
31              (id, file_pdf_name, file_pdf, experiment_name)
32              VALUES (?, ?, ?, ?)"""
33
34          # define data to be inserted as tuple
35          data_tuple = (
36              experiment_id,
37              pdf_name,
38              convert_to_binary_data(
39                  Path.cwd() / "for_upload" / pdf_name
40                  ),
41              experiment_name
42              )
43
44          # execute command and commit
45          cursor.execute(sqlite_insert_blob_query, data_tuple)
46          sqlite_connection.commit()
47
48          print("Image_and_file_inserted_successfully_as_a_BLOB_into_\
49  table.")
50          cursor.close()
51
52      except sqlite3.Error as error:
53          print("Failed_to_insert_blob_data_into_sqlite_table.", error)
54
55      finally:
56          if sqlite_connection:
57              sqlite_connection.close()
58              print("The_SQLite_connection_is_closed.")
59
60
61  # do insertions
62  insert(1, "neutral_result", "A.pdf")
63  insert(2, "positive_result", "B.pdf")
```

Discussion

In the above example, it is important to understand, that there is no processing involved in the sense of an extract transform load (ETL)-pipeline. The main focus is to store data obtained experimentally "as is". But why should this be relevant to your daily work? First and foremost, it is a way to ensure flexibility for posterior analysis. Assuming you have access to an appropriate file parser, you'll have the option of focusing on different aspects of the files' content that you may not be currently aware of. You might be interested in the experimental data today, but in the metadata (What day was the measurement taken? Which user performed the measurement?) under different circumstances, especially when unexpected results occur. It is also a way to foster collaboration within research institutes or industrial research. Since researchers may have different perspectives or hypotheses about a particular effect or observation, they also might want to look at the same data in a (slightly) different way. Classical examples include the position of broad peaks in spectroscopy or the onset time, i. e., the time at which a *certain* slope is exceed in thermal analysis. Frequently, these slightly different ways of looking at the same set of data lead to the creation of numerous methods or SOPs in highly structured organizations. Due to the sheer number of them, maintaining them entails a considerable – probably gratuitously excessive – effort. Having the raw data available, each researcher can take a look at it according to his or her expertise and may find out that the differences from other points of view are smaller than expected.

Further reading

Often, working on a single "flat" table is sufficient for collecting only scientific data and information corresponding to individual samples to be analyzed. In situations where experimental results related to a sample should be connected to information about its composition, relationships between the individual tables have to be defined. The key ideas of building a relational database are summarized on https://www.sqlitetutorial.net/sqlite-create-table/.

24 Reading from a database

Problem

You want to select *data* and *information* stored in a database associated with certain objects for further analysis. This can be considered the reverse way of the process described in Concept 23.

Context

Having "everything" in one place and easily accessible is one of the first and most important steps in conducting experimental research and being successful at it. Necessarily, "everything" relates exclusively to all the data and information collected in advance either experimentally or from other reliable sources.

To be more specific, we might be interested in all the experimental raw data collected using *method X* for all the samples within a particular project for further analysis. This further analysis could be – depending on the type of raw data – something like the identification of peak positions, calculation of integrals within a certain range, but also as easy as selecting defined values from a list. In this sense, the options are almost endless and mostly limited by the types of analyses you have in mind and the set of "conventionally applied" analyses within a given scientific domain. The key idea here is the following: By having access to the actual experimental raw data as obtained from the respective device (as xlsx, pdf, txt, csv, other proprietary formats), you'll be able to ask and also answer questions retroactively, despite them *not* being the initial focus of your data collection and analysis efforts.

Storing experimental raw data files along with other information related to a formulation or reaction batch is one way of ensuring retrospective flexibility.

Solution

The first approach to retrieving experimental raw data from the previously generated *example.db* relies on the sqlite3-package. After defining the path to which experimental data is retrieved from the database, the helper function write_to_file is defined in code section 24.1. It takes the binary data as stored in the database and writes it to a local file. From there, it is available for conventional further processing using scripts. The defined read-function takes the experiment's identification number as an argument and saves the file to the local folder taking into account the information from within the database.

https://doi.org/10.1515/9783111334608-028

Listing 24.1: Reading from the SQLite-database *example.db*.

```python
import sqlite3
from pathlib import Path

# define target path
target_path = Path.cwd() / "from_download"
# create directory if it does not exist
if not target_path.exists():
    # create
    target_path.mkdir()


def write_to_file(data, filename):
    """
    converts binary data to proper format and writes it to filename.
    """
    # Convert binary data to proper format and write it on Hard Disk
    with open(filename, "wb") as file:
        # write data
        file.write(data)

    # info
    print("Stored_BLOB_data_into:_", filename, "\n")


# # target table
# '''CREATE TABLE experiments (
#      id INTEGER PRIMARY KEY,
#      file_pdf_name TEXT NOT NULL,
#      file_pdf BLOB NOT NULL,
#      experiment_name TEXT NOT NULL
#      );'''

def read(experiment_id):
    """
    read database contents related to "experiment_id"
    """
    try:
        sqlite_connection = sqlite3.connect("example.db")
        cursor = sqlite_connection.cursor()
        print("Connected_to_SQLite")

        # define query
        sqlite_fetch_blob_query = """SELECT * FROM experiments
            WHERE id = ?"""

        # fetch query
        cursor.execute(sqlite_fetch_blob_query, (experiment_id,))

        # get record
        record = cursor.fetchall()

        for row in record:
            print(f"id___:_{row[0]}")
            print(f"name_:_{row[3]}")
            # info
            print("..._Storing_pdf_on_disk.\n")
            # pdf
```

```
58              print(f"␣␣␣-␣{row[1]}␣found.")
59              write_to_file(row[2], target_path /
60                  f"{experiment_id:03d}_aka_{Path(row[1]).stem}.pdf"
61                  )
62
63          # close
64          cursor.close()
65
66      except sqlite3.Error as error:
67          print("sqlite3.Error:␣", error)
68
69      finally:
70          if sqlite_connection:
71              sqlite_connection.close()
72              print("The␣SQLite␣connection␣is␣closed")
73
74
75  # read data and information from database + storing files locally
76  # for further processing.
77  read(1)
78  read(2)
```

Please note that this code snippet will run successfully only if the code from the previous Concept 23 has been executed and the resulting SQLite database file *example.db* exists.

An alternative approach for reading the database contents and storing the results in local files is demonstrated in code section 24.2. It's main difference is the dependency on the SQLAlchemy package. Accordingly, we build a query to get all the table contents via the variable results and define some printed output in addition to saving the experimental data as local files from within a for loop.

Listing 24.2: Reading *data* and *information* from the SQLite-database *example.db*.

```
1   import sqlalchemy
2   from pathlib import Path
3
4   # define target path
5   target_path = Path.cwd() / "from_download"
6   # create directory if it does not exist
7   if not target_path.exists():
8       # create
9       target_path.mkdir()
10
11  def write_to_file(data, filename):
12      """
13      converts binary data to proper format and writes it to filename.
14      """
15      # Convert binary data to proper format and write it on Hard Disk
16      with open(filename, "wb") as file:
17          # write data
18          file.write(data)
19
20      # info
```

```
21      print(f"Stored_blob_data_into:_{filename}")
22
23   # # target table
24   # '''CREATE TABLE experiments (
25   #       id INTEGER PRIMARY KEY,
26   #       file_pdf_name TEXT NOT NULL,
27   #       file_pdf BLOB NOT NULL,
28   #       experiment_name TEXT NOT NULL
29   #       );'''
30
31   engine = sqlalchemy.create_engine(
32               "sqlite:///example.db",
33               echo=True
34               )
35
36   inspector = sqlalchemy.inspect(engine)
37
38   # info on available table(s)
39   for t in inspector.get_table_names():
40       print(t)
41
42   # info on available columns within table experiments
43   for idx, c in enumerate(inspector.get_columns("experiments")):
44       print(f"{idx}_:_{c['name']}")
45
46   # query statement
47   query = "SELECT_*_FROM_experiments"
48   # get all results from query
49   results = engine.connect().execute(sqlalchemy.text(query)).fetchall()
50
51   # loop result rows
52   for result in results:
53       # indicator
54       print("\n=======================")
55       for idx, (col, entry) in enumerate(zip(
56               inspector.get_columns("experiments"),
57               result
58               )):
59          # print-output if "non-BLOB"
60          if idx in [0, 1, 3]:
61              print(f"{idx}_|_{col['name']:<15}_:_{entry}")
62          # save to file if BLOB
63          else:
64              # info
65              print(f"{idx}_|_____>>_binary_content_<<")
66              # write BLOB to file
67              write_to_file(
68                  entry,
69                  target_path / f"{result[0]:03d}_aka_{result[idx-1]}"
70                  )
71
72   # close engine
73   engine.dispose()
```

Discussion

One of the benefits of SQLAlchemy comes in the form of its many convenience functions. If you are trying to figure out the structure of a table within a database, the functions get_table_names and get_columns are there to help. Also, the corresponding return values can be easily converted to pd.DataFrames for further inspection.

Most importantly, SQLAlchemy is not limited to SQLite-databases. It also allows for connecting to other dialects.[1] All dialects require the installation of an appropriate DataBase Application Programming Interface (DBAPI) driver. Included dialects are PostgreSQL, MySQL, MariaDB, SQLite, Oracle and Microsoft SQL Server.

Further reading

The SQLAlchemy package provides certain more advanced query functionality as shown above as its goal is none other than to change the way you think about databases and Structured Query Language (SQL).[2] It's key features are summarized on https://www.sqlalchemy.org/features.html.

---

1 for a list of supported dialects see here https://docs.sqlalchemy.org/en/20/dialects/index.html
2 https://www.sqlalchemy.org/philosophy.html

Planning experiments and/or building on legacy data/information

Now that you've covered the technical or manual parts of setting up Python appropriately and interfacing with the most common data formats you might encounter in your scientific practice, it's time to turn to science. The following concepts will cover

- how to make existing experimental, i. e., data and information accessible to the programmatic/scripting approach described here by rearranging files according to a defined structure.
- how design of experiments (DOE) tools for planning experiments can support the execution of the "right" experiments.
- how already existing experiments can be used hand in hand with planned experiments according to the DOE approach. Here, it is crucial to understand that the support is mutual: When designing new experiments using a systematic approach you can take into account what has already been done experimentally and what does not work, i. e., which composition or parameter ranges to avoid. If you want to stick to the classical way of creating experiments, getting some inspiration from the systematic DOE-approach might lead to surprising formulations/compositions you hadn't thought of before. Just give it a try.

https://doi.org/10.1515/9783111334608-029

25 Leveraging existing experiments

Problem

You want to make data and information related to a particular project easily accessible for further processing.

Context

Often during your academic or industrial career, you will become part of a project only after some data has been collected already. In an academic setting, you may be asked to continue the work of a former student who "has already done something in this direction". Similarly, in the case of industrial research, you may be working on a particular spin-off project for which data exists from a parent project. More frequently, however, there are personnel changes at the project level, which may require you to quickly get an overview of past activities based on the available data.

Solution

To date, most organizations still rely on an organized folder structure approach to handling data related to the experiments conducted. Within the example used throughout the remainder of this book, we want to take a closer look at a formulation consisting of three components, A, B and C. After mixing these components according to a defined protocol, the resulting material is subjected to different types of characterization. An overview of the experimental procedure is shown in Figure 25.1.

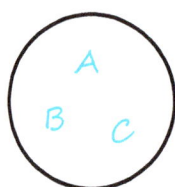

A
B C
Formulation

Calorimetry

Vicat testing

Application tests

Figure 25.1: Formulation example considered throughout the following chapters. The formulations consist of the components A, B and C, and are characterized by several methods.

In order to programmatically check the completeness of the entire experimental set, it is beneficial – in the absence of a database solution acting as a single source of truth – to have same type of data sorted within individual folders. As shown in Figure 25.2, the use of mono-fractional folders containing the respective results is recommended. In the

https://doi.org/10.1515/9783111334608-030

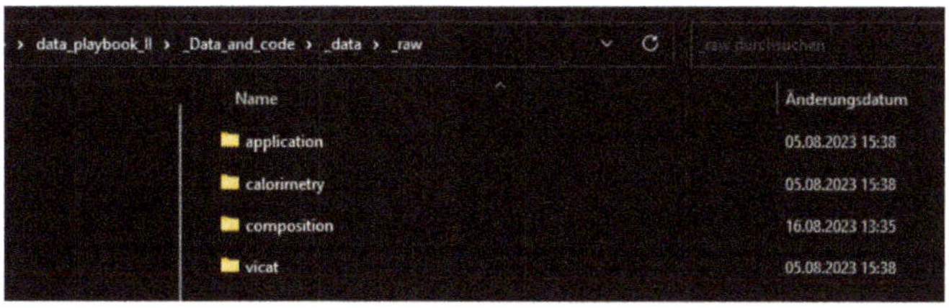

Figure 25.2: Folder structure for separate access to experimental data according to the different methods involved.

example, there's one for the experimental compositions and one for each characterization method.

The following lines of code (see code section 25.1) parse the available formulations from a Microsoft® Excel® file. The following steps check the availability of the corresponding experimental characterizations. Most importantly, the identification number of the formulations defined in the compositions file must to be picked up in the other characterization types to allow for a mapping.

Listing 25.1: Reading experimental raw data from the above ordered folder structure.

```python
import pathlib
import pandas as pd

# define base path to data
data_path = pathlib.Path().cwd().parent.parent /\
    "_Data_and_code" / "_data" / "_raw"

# get composition
compositions = pd.read_excel(
    data_path / "composition" / "composition.xlsx"
    )

# get Vicat characterizations
data_vicat = pd.read_csv(data_path / "vicat" / "vicat.csv")

# get application data
data_application = pd.read_json(
    data_path / "application" / "application.json"
    )

# get list of unprocessed calorimetry files
data_calorimetry = pd.Series(
    i.name for i in (data_path / "calorimetry").iterdir()
    )


# build summary table
overview = compositions[["id"]].reset_index()
# check for corresponding experimental findings
overview["calorimetry"] = [
    True if data_calorimetry.str.match(i).any()
    else False for i in overview["id"]]
```

```
33  overview["vicat"] = [
34      True if data_vicat["id"].str.match(i).any()
35      else False for i in overview["id"]]
36  overview["application"] = [
37      True if data_application["id"].str.match(i).any()
38      else False for i in overview["id"]]
39
40  # get summary
41  overview.sum()
42  # index                                              1540
43  # id            P23_X_001P23_X_002P23_X_003P23_X_004P23_X_005P...
44  # calorimetry                                          56
45  # vicat                                                56
46  # application                                          56
```

As the overview variable shows, there's calorimetry, Vicat and application data for each of the formulations, whose composition is specified in the *composition.xlsx* Microsoft® Excel® file.

Discussion

Getting an overview of this kind of high-level quality is particularly relevant when you are getting started on new projects. In some cases, it may be necessary to reorder and rename the files to allow for an easy connection, i. e., and to define relationships between different types of experiments and compositions.

Keep in mind to involve your colleagues who may be working on the same project if applicable.

The way data is stored and how files are named can become a source of conflict as some people prefer "speaking names" even for the experimental results file. Instead of just using *P23_X_001* as the unique identifier, they might prefer something like *A52_B19_C29*. Be open to compromise here: A combined approach of using both the identifier for easy machine mapping *and* information for enhanced human accessibility might be a solution. The previous example could be coined as *P23_X_001__A52_B19_C29*.

Further reading

For a more detailed discussion of the organized folder approach, see *Data Management for Natural Scientists*.[1]

---

1 https://www.degruyter.com/document/doi/10.1515/9783110788433/html

26 Planning experiments

Problem

You want to plan a set of possibly few experiments to be performed, but at the same time ensure maximum information gain.

Context

DOE is defined as a branch of applied statistics concerned with the planning, conducting, analysis, and interpretation of controlled tests to evaluate the factors that control the value of a parameter or group of parameters.[1] From an intuitive point of view, the DOE approach is not natural to most people trained as natural scientists, who regularly prefer the use of "ladder studies". This is evident when comparing DOE with the classical one-variable-at-a-time (OFAT) approach outlined in Figure 26.1.

Taking the very basic example of varying only two parameters, let's say the reaction time and reaction temperature of a chemical reaction, OFAT suggests keeping all but one parameter constant. In this scenario, any changes in the "results end", e. g., experimental yield can be fully attributed to that one specific parameter that was changed. One the one hand, this has a positive aspect in terms of an exclusive attribution of *cause to effect*.[2] On the other hand, combined or interaction effects are systematically discarded in the OFAT-approach. But just because we do not look at them in the chosen set of experiments does not mean that they are not there.[3]

The DOE approach, however, aims to vary multiple parameters simultaneously according to a systematic pattern. This necessarily involves a sense of less control on the part of the researcher or technician carrying out the actual experiments. Interaction effects become more important than mono-causal relationships. But DOE also has some positive aspects:

- Having to perform fewer experiments to gain a comparable amount of insight into a given system compared to the OFAT approach.
- The ability to work with complex real-world formulations rather than simplified model systems, as you can easily vary five to eight parameters at a time.

---

1 https://asq.org/quality-resources/design-of-experiments

2 That's what scientists are looking for!

3 A classic – perhaps oversimplified – example of this type of interaction of cause and effect is the sweetening of coffee. Assume you have a cup of coffee. The parameters for modifying the system are 1) adding a spoonful of sugar and 2) stirring. Adding sugar alone and stirring alone will not change the sweetness of your coffee. Only the combination will. Using DOE, you are at least invited to perform this kind of interaction experiments, which can make all the difference in the world.

https://doi.org/10.1515/9783111334608-031

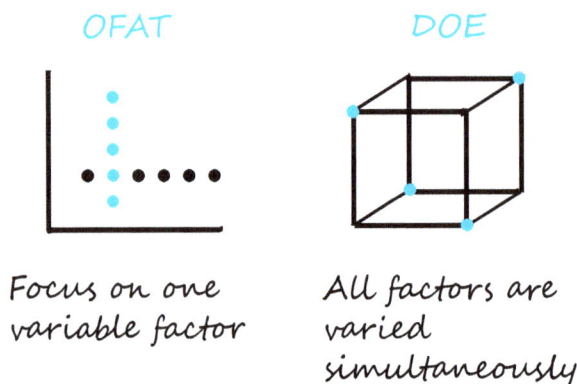

Figure 26.1: The OFAT and DOE approaches compared side by side.

– Being able to make predictions of expected results based on the experimental evidence already collected.

Solution

Use the dexpy package to design experiments using the DOE approach. In our example, we'd like to get proposed compositions for a ternary formulation consisting of components A, B and C. In these kind of mixtures, the relative contents of the components add up to 1, so essentially we are dealing with a two-parameter scenario here.[4] Since dexpy is not intentionally designed to handle such cases, some adaptions to the resulting pd.DataFrames are required. As shown in Figure 26.1 we still receive a list of proposed experiments.

Listing 26.1: Creating a set of proposed experiments for a ternary formulation using dexpy.

```
1  import dexpy.optimal
2  from dexpy.model import ModelOrder
3  import pandas as pd
4  import numpy as np
5
6  # ensure repeatability of experimentes generation
7  np.random.seed(15)
8
9  # define components names and limits to be used in the design
10 column_names = ['A', 'B']
11 actual_lows = {'A': 0.4, 'B': 0.1 }
12 actual_highs = {'A': 0.65, 'B': 0.3 }
13
14 # create the "optimal" design taking varying two factors (A and B)
```

---

4 ... for the sake of easy-to-understand visualizations.

```
15  formulation_design = dexpy.optimal.build_optimal(
16      2,
17      order=ModelOrder.quadratic,
18      run_count=6
19      )
20  # manually define "center points"
21  center_points = [
22      [0.2, 0.2],
```

To visualize the ternary formulations created by dexpy, the mpltern-package is used to enable ternary projections of this kind of compositions.[5] In the scatter plot, each point represents a target experimental composition to be prepared and subsequently analyzed. The position within the triangle represents the composition in terms of the relative amounts of A, B and C.

Listing 26.2: Visualization of the proposed experiments of a ternary formulation as obtained from dexpy.

```
1   # combine the dexpy suggested "optimal" points and manually added
2   # center point to obtain all experiments
3   formulation_design = pd.concat([
4       formulation_design,
5       pd.DataFrame(center_points*2, columns=formulation_design.columns)
6       ])
7
8   # "inform model about settings"; apply "true" column names
9   formulation_design.columns = column_names
10
11  # map "-1 .. 1" range to true values according to variables
12  # "actual_lows" and "actual_highs"
13  actual_design = dexpy.design.coded_to_actual(
14      formulation_design,
15      actual_lows,
16      actual_highs
17      )
18
19  # infer content of third component "C" (relative contents of
20  # components have to add up to 1)
21  actual_design["C"] = 1 - actual_design["A"] - actual_design["B"]
22
23  # discard "illegal ones" (only contents of A, B and C>=0 are allowed)
24  actual_design = actual_design.query("C >= 0")
25
26  # reset index to allow for appropriate color mapping
27  actual_design = actual_design.reset_index(drop=True)
28
29  # info
30  print(actual_design)
31
32  # save design to Excel
33  actual_design.to_excel("design.xlsx")
34
```

5 see https://mpltern.readthedocs.io/en/latest/

```
35  # %% vizualization
36
37  import matplotlib.pyplot as plt
38  from matplotlib import colors
39  import mpltern  # allows to use "projection="ternary"
40
41  # copy formulation design data
42  data = actual_design.copy()
43
44  # define figure
45  fig = plt.figure()
46  # initialize ternary plot
47  ax = fig.add_subplot(projection="ternary")
48
49  # define settings for discrete colorbar
50  cmap = plt.cm.viridis
51  norm = colors.BoundaryNorm(np.arange(-0.5, len(data)+0.5, 1), cmap.N)
52
53  # indicate compositions of formulations as scattered dots in
54  # ternary diagram
55  pc = ax.scatter(
56          data.A,
57          data.B,
58          data.C,
59          c=data.index,
60          cmap=cmap, norm=norm
61          )
62
63  # highlight ranges
64  ax.axtspan(actual_lows["A"], actual_highs["A"], fc="k", alpha=0.2)
65  ax.axlspan(actual_lows["B"], actual_highs["B"], fc="k", alpha=0.2)
66
67  # add "equal ratio line"
68  ax.axline(
69      [0.0, 1.0, 0.0],
70      [0.4, 0.0, 0.6],
71      color="k",
72      )
73  # add info text
74  ax.text(0.3, 0.3, 0.4, 'equal_ratio_A:C__', ha='right',
75          va='center', size=8)
76
77  # set labels
78  ax.set_tlabel('A', size=20)
79  ax.set_llabel('B', size=20)
80  ax.set_rlabel('C', size=20)
81
82  # define title
83  ax.set_title("$dexpy$_Suggested_Experiments")
84
85  # define axis for color bar axis
86  cax = ax.inset_axes([1.05, 0.1, 0.05, 0.95], transform=ax.transAxes)
87  colorbar = fig.colorbar(pc, cax=cax, ticks=range(len(data)))
88  colorbar.set_label("Experiment_#", rotation=270, va="baseline")
89
90  # save figure
91  plt.savefig("dexpy_experiments.png", dpi=300, bbox_inches="tight",
92              transparent=True)
```

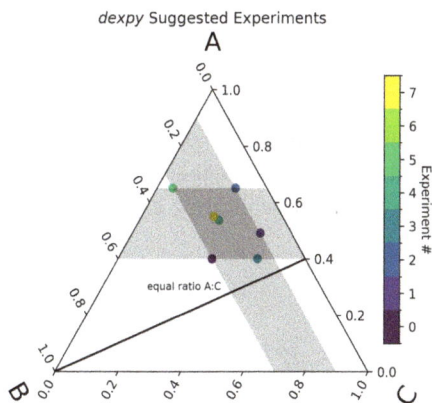

Figure 26.2: Experiments suggested by the dexpy-package for a formulation consisting of components A, B and C taking into account user-defined boundaries displayed in a conventionally used ternary phase diagram.

The resulting visualization of the experiments to be performed after the processing outlined above is given in Figure 26.2.

A primer on ternary diagrams: ℹ️
–   The tips of the triangles mark the pure compounds A, B and C.
–   On each baseline, the compound noted on the opposite tip is absent. On these baselines there is a gradual transition of the corresponding binary mixtures.
–   The closer a point is to a tip of the triangle, the higher is the content of that compound.

On the added *equal ratio A:C*-line, the ratio of A to C remains constant at 0.4:0.6, i. e., 2:3. At the upper right side of the line, the mixture is binary, i. e., consisting of two components, while its pure compound B at the lower left end. In between, increasing amounts of component B are added to the initial binary A:C.

Discussion

DOE is certainly an approach to (industrial) science that takes some getting used to. Depending on the field you are working in, it's either more or less important to conduct as few experiments as possible to ensure maximum information gain. In a production-type setting, where one experiment means occupying, say, six people's time for a day and producing 10 tons of non-marketable product, you would definitely think a bit more about the necessity than you would in a conventional R&D-type setting, where producing one more formulation might take 15 minutes of one person's time and a negligible amount of additional raw material cost.

In this sense, as R&D scientists, we are in a comfortable position to use DOE as an additional source of inspiration or list of suggestions for the set of experiments to be conducted. You'll also be able to discard certain suggested experiments, thereby sacrificing some predictive power of the resulting model, but also to add other experiments not foreseen by your DOE-tool of choice. After all, the subsequent analysis will benefit from the additional data points. This is especially true for replicating experiments in order to get an estimate of repeatability.

Further reading

5 Further Python-packages supporting DOE are DoEgen[6] and doepy.[7]

There are also a number of commercial software solutions that use the DOE approach available. Among the most popular in industrial research are JMP[8] and MODDE.[9] The strength of these software packages is certainly that they allow much more fine-tuning in a highly user-friendly way via sophisticated Graphical User Interfaces (GUIs).

A detailed overview of the tools required for optimal experimental design and analysis is given in the *Handbook of Design and Analysis of Experiments.*[10]

---

6 https://pypi.org/project/DoEgen/

7 https://pypi.org/project/doepy/

8 https://www.jmp.com/en_us/home.html

9 https://www.sartorius.com/en/products/process-analytical-technology/data-analytics-software/doe-software/modde

10 https://www.routledge.com/Handbook-of-Design-and-Analysis-of-Experiments/Dean-Morris-Stufken-Bingham/p/book/9780367570415

27 Using legacy and planned experiments hand in hand

Problem

You want to gain additional insight based on (or "on top of") already existing characterization results.

Context

Continuing on the case outlined in Concept 26, where formulation compositions for ternary mixtures consisting of components A, B and C were obtained using the dexpy-package, the following situation might arise: Let's assume, one of the previously created mixtures was identified as meeting a first set of "technical" criteria. After some test runs with potential customers, it's discovered that further modifications are needed to meet certain – previously unknown – constraints in terms of color and storage stability. Based on existing expertise, a colleague decides to add two more compounds, X and Y, in small amounts to counteract these undesired effects, adding an additional layer of complexity. In essence, the former ternary formulation becomes a mixture of five components, as outlined in Figure 27.1.

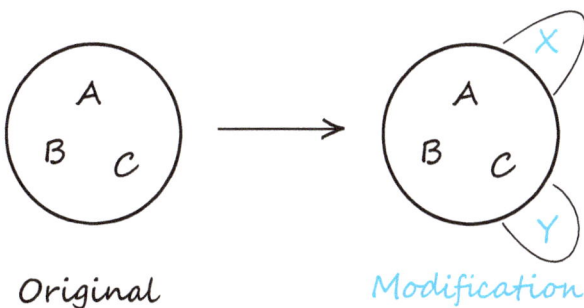

Figure 27.1: Conceptual sketch of the creation of new formulations based on existing compositions for which some experimental data are available.

But does this mean that the previous experiments on the ternary system have no value for the updated question? Of course not. The ternary mixtures can be considered as the "trivial cases" of five-component mixtures, where the relative content of components X and Y is set to zero. From this perspective, we have been collecting experiments for the five-component formulations all along. The challenge is to translate this understanding into the following lines of code.

https://doi.org/10.1515/9783111334608-032

Solution

The key to understanding the five-component system based on the experiments already conducted for the ternary mixtures is based on
- "rewriting" the legacy experiments in terms of the new "broader" experiments.
- generating a new set of suggested experiments via a DOE package. Again, we use dexpy (as introduced in Concept 26).
- sorting out suggested experiments that are considered irrelevant or duplicates of the existing experiment based on a similarity criterion.

The suggested procedures are shown in code section 27.1. An important step is to determine how "different" two experiments are from each other in terms of their composition. This is where *distance metrics* comes in. This term refers to functions that determine a value to describe the pairwise difference between observations in an *n*-dimensional space as a non-negative real number. In simpler terms, it allows us to answer questions of the type: *How different is formulation number 4 from formulation number 8 expressed as a single number? Is this difference greater or smaller than for formulations number 5 and 16?*

Listing 27.1: Upgrading legacy three-component formulations to five-component formulations via dexpy.

```python
import pandas as pd
import numpy as np
import dexpy
from scipy.spatial.distance import pdist, squareform

# parse available ternary mixtures
compositions_3 = pd.read_excel("design.xlsx", usecols="B:D")

# "upgrade" three component to five component formulation
compositions_5 = compositions_3.copy()
compositions_5[["X", "Y"]] = 0

# determine concentration ranges taking into account maximum
# additions of components X and Y
compositions_5_max = compositions_5.copy()
# assume maximum addition of X to be 0.1
compositions_5_max["X"] = 0.1
# assume maximum addition of X to be 0.2
compositions_5_max["Y"] = 0.2

# within a "formulation", relative contents add up to 1
compositions_5_max = compositions_5_max.div(
    compositions_5_max.sum(axis=1), axis=0
)

# "describe" rescaled formulations to get inputs for DOE
limits = compositions_5_max.describe().loc[["min", "max"],:]

# ensure repeatability of experimentes generation
np.random.seed(15)
```

```python
32  # define components names and limits to be used in the design
33  column_names = ['A', 'B', 'X', 'Y']
34  actual_lows = {
35      'A': limits.at["min", "A"],
36      'B': limits.at["min", "B"],
37      'Y' : 0,
38      'X' : 0
39      }
40  actual_highs = {
41      'A': limits.at["max", "A"],
42      'B': limits.at["max", "B"],
43      'Y' :limits.at["max", "X"],
44      'X' : limits.at["max", "Y"],
45      }
46
47  # create the "optimal" design taking varying two factors (A and B)
48  formulation_design = dexpy.optimal.build_optimal(
49      len(column_names),
50      order=dexpy.model.ModelOrder.quadratic,
51      run_count=15
52      )
53
54  # "inform model about settings"; apply "true" column names
55  formulation_design.columns = column_names
56
57  # map "-1 .. 1" range to true values according to variables
58  # "actual_lows" and "actual_highs"
59  proposed_design = dexpy.design.coded_to_actual(
60    formulation_design,
61    actual_lows,
62    actual_highs
63    )
64
65  # infer content of third component "C" (relative contents of
66  # components have to add up to 1)
67  proposed_design["C"] = 1 - proposed_design["A"] -\
68      proposed_design["B"] - proposed_design["X"] -\
69      proposed_design["Y"]
70
71  # discard "illegal ones" (only contents of A, B and C >= 0
72  # are allowed)
73  proposed_design = proposed_design.query("C >= 0")
74
75  # reset index to allow for appropriate color mapping
76  proposed_design = proposed_design.reset_index(drop=True)
77
78  # info
79  print(proposed_design)
80
81  # combine proposed design ans already existing 3 component design
82  overall_design = pd.concat([
83      compositions_5,
84      proposed_design
85      ]).reset_index(drop=True)
86
87  # info
88  print(overall_design)
89
90  # save results
91  overall_design.to_excel("design_5_components.xlsx", index=False)
```

```
92
93   # calculate euclidian difference between individual experiments?
94   # how much are they distinct?
95   distances = pdist(overall_design.values, metric='euclidean')
96   dist_matrix = pd.DataFrame(squareform(distances))
97
98   # restrict to most distinct
99   helper = dist_matrix.iloc[:8,8:][dist_matrix > 0.45].\
100      dropna(axis=1, how="all")
101
102  # info
103  print(
104      "Most_distinct_experiments_to_legacy_experiments:",
105      ",_".join([str(i) for i in helper.columns])
106      )
```

---

i Take a look at the dist_matrix variable and compare the differences between the initial ternary formulations and the top five component formulations built on it.[1]

Discussion

The "mixture of mixtures" scenario described here is quite common in an industrial research type setting. Many times, formulation competence, i. e., which and how these components are mixed, is the key to success and failure of entire companies.[2]

This kind of game can obviously also be played the other way around. A formulation project could start with mixing, e. g., a solvent, an antioxidant, and ten different types of hardeners as a screening study. To allow for a simplistic manufacturing, a process requirement might be to limit the number of hardeners used in a formulation to a certain threshold. Assuming that the composition of the screened hardener components is available to you (to a "sufficiently precise level"; this is where domain and expert knowledge comes into play) either from product information or analyses, the experiments conducted could be rewritten: In this scenario, the number of compounds within the final formulations can be reduced assuming that the screened hardeners consist themselves of the same components in varying relative proportions. In this scenario, going one level "deeper", helped to reduce the apparent complexity of the set of experiments. In a sense, it is the task of a scientist to find a representation of experiments that is as simple as possible so that maximum insight can be extracted.

---

1 Use, e. g., the command dist_matrix.head(8).median().plot(kind="bar").
2 Just recall the recall in the infamous recipe of Coca Cola.

Further reading

In addition to building experimental designs with numerical options, the dexpy package also allows you to take into account categorical data.[3]

---

[3] https://statease.github.io/dexpy/example-coffee.html

Collecting experimental data / lab work phase

Slowly but steadily we are approaching the realm of "real" hands-on lab work. Aside from the preparative aspects, we want to focus on more data-intense tasks associated with many characterization methods such as spectroscopy, thermal analysis, and wet chemical analysis, to name only a few. Ultimately, you'll end up with either an individual results file or an entry into a database system. This may be facilitated by the software provided by the equipment manufacturer, a solution provided by your research institute or company, or a licensed third-party solution. Either way, in order to bring in your domain and expert knowledge on the actually obtained experimental data, you need to be able to "work" with it.

The ability to simply use already existing solutions, i. e., Python packages to handle dedicated tasks is certainly the more convenient situation (see Concept 28).

If you are not aware of an existing solution and cannot find a suitable one on either the public or private package indices available to you, you have no choice but to build the solution you need to perform your desired workflow in an efficient and – most importantly – reproducible manner (see Concept 29).

Both scenarios are described below.

https://doi.org/10.1515/9783111334608-033

28 Using dedicated modules – use what's available

Problem

You have collected a set of filed-based experimental raw data from a particular type of analytic instrument and have a specific type of analysis in mind. The software package provided by the manufacturer does not allow for scientific analysis of the generated results but is designed only to control machine settings and experimental procedures.

Context

Much to the surprise and astonishment of numerous scientists and analyticas practitioners, the standard software that comes with analytical instruments regularly does not allow for the extraction of user-specified characteristics from "typical" results. Of course, the aforementioned typical shape may vary more or less strongly depending on the application for which a vendor sells its instrument. Nevertheless, this remains a source of considerable workarounds and manual effort on many sides. Even when a certain type of analysis is provided by the software and intended by the manufacturer, the possibility to compare these parameters across a set of experiments (such as the set of formulations created in Concept 26) in a reasonable, user-friendly, and appealing way is all too often not available out of the box.

To overcome this lack of analysis and visualization capabilities, most researchers rely on a combination of considerable manual effort and more or less developed Microsoft® Excel® skills.

Solution

Navigate to https://pypi.org/ and try to identify a package that meets your requirements based on the respective descriptions and the keywords provided. Assuming we have collected experimental heat flow calorimetry results,[1] appropriate search terms to provide might be the name of the method and the manufacturer of the device. A screenshot of the The Python Package Index (PyPI)'s landing page is given in Figure 28.1.

When you run the search, you'll be provided with a list of packages sorted by relevance, as the engine assumes. Be prepared to *not* find the solution you'll end up working with as the top results (see Figure 28.2).

---

1 For example using tools from https://www.tainstruments.com/new-tam-air/

https://doi.org/10.1515/9783111334608-034

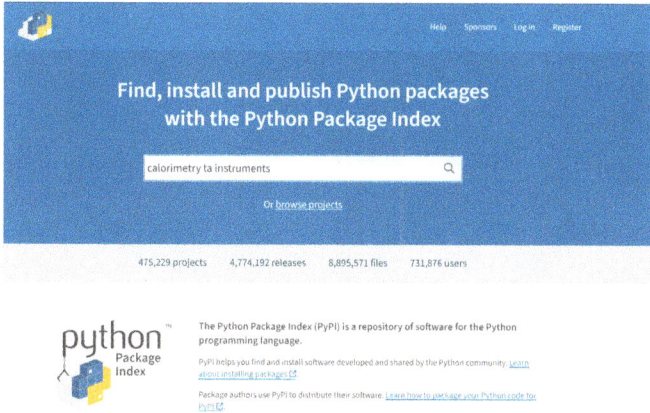

Figure 28.1: Screenshot of PyPI's landing page.

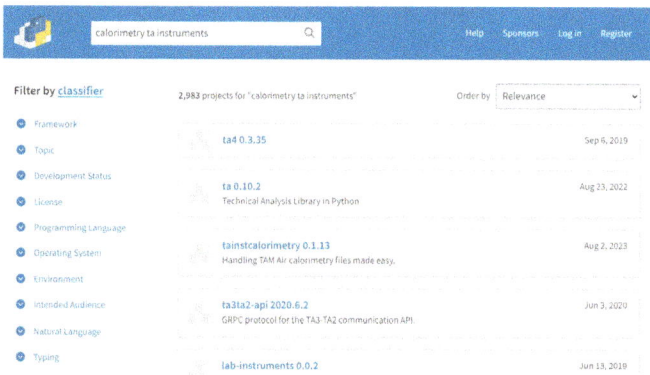

Figure 28.2: Keyword search on PyPI.

Depending on how well the package matches the keywords you are looking for, it will appear higher or lower in the list. Also, consider installing multiple packages aiming to solve the task you have in mind, and follow the – hopefully provided – examples that come with the package of your choice. In this example, we'll continue with the `tainstcalorimetry` package, whose landing page is shown in Figure 28.3.

Following the example usage suggested there, the first step is to define the local path to the experimental raw data. After the results have been parsed, they are available for further plotting and analysis tasks as outlined in code section 28.1.

Listing 28.1: Using `tainstcalorimetry` for basic plotting and analysis.

```
1  import TAInstCalorimetry.tacalorimetry as ta
2
3  # intialialize "Measurement" object
```

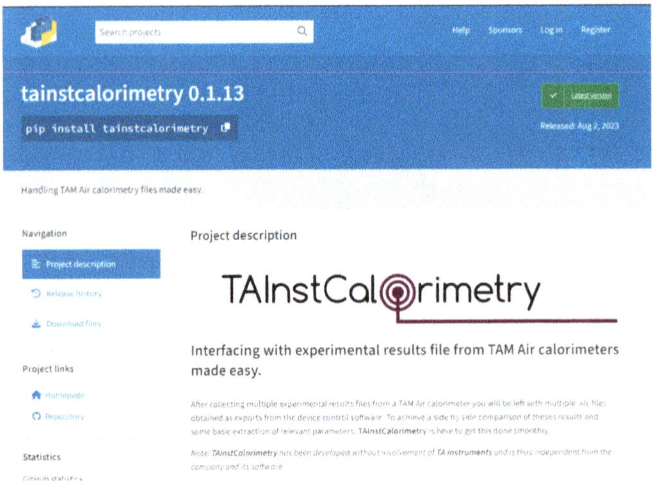

Figure 28.3: Homepage of the `tainstcalorimetry` package on PyPI.

```
4   measurements = ta.Measurement(
5       folder="data", show_info=True, auto_clean=False
6       )
7
8   # plot
9   measurements.plot()
10  # set axes limits in plot
11  ta.plt.ylim(0, 3)
12  ta.plt.xlim(0, 48)
13  # show plot
14  ta.plt.show()
15
16  # get peaks
17  peaks = measurements.get_peaks(
18      show_plot=False,
19      cutoff_min=60,
20      )
21
22  # plot
23  measurements.plot()
24  # get axes for further modification
25  ax = ta.plt.gca()
26
27  # set axis limtis
28  ax.set_ylim(0,3)
29  ax.set_xlim(0, 15)
30
31  # loop plots and get corresponding peaks
32  for line in ax.lines:
33      # get label
34      label = line.get_label()
35      # get color
36      color = line.get_color()
37      # plot vertical incator line in matching color
38      ax.axvline(
39          float(peaks[peaks["sample_short"] == label]["time_s"])/3600,
```

```
40              color=color,
41              linewidth=0.5
42          )
43
44  # save plot
45  ta.plt.savefig("plot.png", dpi=300, bbox_inches="tight")
```

Basic plotting allows you to compare the collected experimental data in just a few lines of code. The same is true for the evaluation of peak positions. Most of the code above is actually dedicated to specifying the plots according to the idea in mind: side-by-side comparison of the samples with an indication of the identified peak positions (see Figure 28.4).

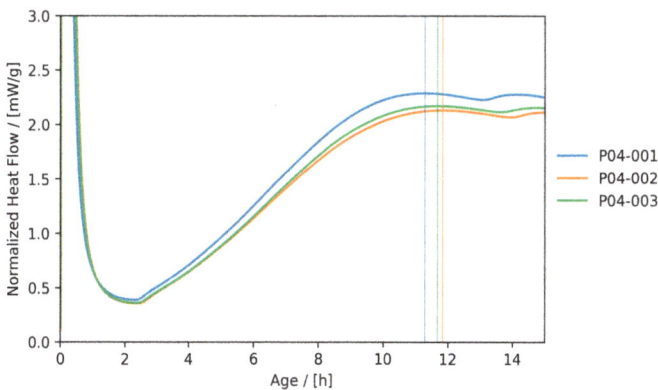

Figure 28.4: Representation of experimental *data* and extracted peak position *information* based on the tainstcalorimetry package.

Discussion

Since these "helper" packages are – at the time of writing – provided by researchers rather than the actual instrument manufacturers, it is quite common for multiple packages to be available on PyPI for the same analytical instrument. Therefore, it may take some effort on your part to identify the one that suits your needs best. Also, don't be afraid to contact the maintainers listed on the projects landing pages on PyPI directly (unless they explicitly write on the project pages that they do not wish to be contacted). Many of them will be happy about further input and will appreciate the support of additional contributors.

Further reading

Available packages related to common analysis devices include:

- `aprheology`[2] allows processing of experimental results from Anton Paar rheometers.[3]
- `bruker-utils`[4] is a module of functions for easy handling of Bruker NMR data in Python.
- `tainstcalorimetry`[5] allows interfacing with experimental results file from TAM Air calorimeters.

---

2 https://pypi.org/project/aprheology/

3 https://www.anton-paar.com/us-en/products/group/rheometers/

4 https://pypi.org/project/bruker-utils/

5 https://pypi.org/project/tainstcalorimetry/

29 Using dedicated modules – build what's missing

Problem

You have collected a set of file-based experimental raw data from a particular type of analytical device and have a certain type of analysis in mind. The software package provided by the vendor does not allow scientific analysis of the generated results *and* (as opposed to Concept 28) you cannot find a Python package that provides a sufficient level of convenience for handling everyday tasks related to this type of experimental raw data.

Context

Much to the surprise and astonishment of numerous scientists and practitioners of analytical tasks, the standard software that comes with analytical equipment regularly does not allow the extraction of user-specified characteristics from "typical" results. Of course, the aforementioned typical shape may vary more or less strongly depending on the application for which a vendor sells its instrument. Nevertheless, this remains a source of considerable workarounds and manual effort on many sides. Even when a certain type of analysis is provided for by the software and intended by the manufacturer, the ability to compare these parameters across a set of experiments (such as the set of formulations created in Concept 26) in a reasonable user-friendly, and appealing way is all too often not achievable out of the box.

To overcome this lack of analysis and visualization capabilities, most researchers rely on a combination of manual effort and more or less developed Microsoft® Excel® skills.

Solution

Combine the capabilities coming with already existing Python packages for your specific purposes to achieve your desired "what you get is what you mean" capabilities. In terms of natural scientific data, these tasks can often be reduced to the following steps:
- extracting tabular experimental data to a pd.DataFrame
- extracting, i. e., picking characteristic values from this pd.DataFrame and/or
- extracting further parameters from the pd.DataFrame that are not accessible from it alone, but rely on more dedicated analysis using, e. g., the scipy and/or lmfit packages.

The code section 29.1 introduces scripts for the tasks of "getting data" and "getting peaks" based on artificially created result files.

https://doi.org/10.1515/9783111334608-035

Listing 29.1: Building a script as a first step towards the get_data function.

```python
import pathlib
import pandas as pd
import matplotlib.pyplot as plt

# path to file
data_path = pathlib.Path().cwd() / "calorimetry"

# pick one file
data_file = next(data_path.iterdir())

# parse file
data = pd.read_excel(data_file)

# plot
data.plot(x="time_s")
```

The next step according to code section 29.2 is defining the get_data and the get_peaks functions.

Listing 29.2: Defining the get_data and the get_peaks functions.

```python
# convert script to function
def get_data(file : pathlib.Path, show_plot : bool = True):
    """
    read experimental heat flow calorimetry data form file and
    return as pd.DataFrame

    Parameters
    ----------
    file : pathlib.Path
        DESCRIPTION.
    show_plot : bool, optional
        DESCRIPTION. The default is True.

    Returns
    -------
    data : TYPE
        DESCRIPTION.

    """
    # info
    print(file.name)

    # parse file
    data = pd.read_excel(file)

    if show_plot:
        # plot
        data.plot(x="time_s", label=file.name)

    # return
    return data
```

```
34  def get_peak(file : pathlib.Path, show_plot : bool = True):
35      """
36      read experimental heat flow calorimetry data form file and
37      return indication for peak position.
38
39      Parameters
40      ----------
41      file : pathlib.Path
42          DESCRIPTION.
43      show_plot : bool, optional
44          DESCRIPTION. The default is True.
45
46      Returns
47      -------
48      TYPE
49          DESCRIPTION.
50      TYPE
51          DESCRIPTION.
52
53      """
54
55      # get data
56      data = get_data(file)
57
58      # add file name to peak
59      data["file"] = file
60
61      # get index corresponding to maximum heeat flow
62      idxmax = data["normalized_heat_flow_w_g"].idxmax()
63
64      if show_plot:
65          # plot
66          data.plot(x="time_s", label=file.name)
67          # add indicator line
68          plt.axvline(data.at[idxmax, "time_s"], color="red", alpha=0.5)
69
70      # return peak "position" a.k.a. time and "maximum row"
71      return data.at[idxmax, "time_s"], data.loc[idxmax,:]
```

To conclude, the defined functions are used in code section 29.3.

Listing 29.3: Using the get_data and the get_peaks functions defined above.

```
1
2   # initialize generator based on the directory
3   files = data_path.iterdir()
4
5   d1 = get_data(next(files))
6   d2 = get_peak(next(files))
7
8   # %% get list of peaks
9
10  # init list
11  peaks_list = []
12
13  # loof files
14  for file in data_path.iterdir():
```

```
15    # info
16    print(file)
17    # get peak info
18    _, this_peak = get_peak(file, show_plot=False)
19    # add to "results list"
20    peaks_list.append(this_peak)
21
22 # compile peaks list to DataFrame
23 peaks = pd.DataFrame(peaks_list)
24
25 # info
26 print(peaks)
```

Discussion

If you take the effort of creating utilities such as the get_data and the get_peaks functions defined above, sharing them within your (trusted) network is highly appreciated (see Concept 45).

---

i To test the functionality of your functions, it is recommended to transfer them to a new module, i. e., file and install it locally. Therefore,
- initialize the new package using poetry poetry new,
- copy the former scripts and transfer them to one or several modules within the package,
- install the package locally pip install -e . and
- optionally, build the package locally for sharing with poetry build.

Further reading

Based on this idea, packages referenced in Concept 28 have been created.

Visualization of experimental results

In order to provide both *data* and *information* for introducing the various plotting options, the following scripts (code section 29.1 and code section 29.2, respectively) have been compiled. They are used to collect *clean* data into Microsoft® Excel® files for easy plotting.

Listing 29.1: Collecting experimental *data* across multiple source files for further processing and visualization.

```python
1  import pathlib
2  import pandas as pd
3  from mycalohelpers import basics
4
5  # data base path
6  path_data = pathlib.Path().cwd().parent / "_data" / "_raw"
7
8  # init list of dataFrames
9  list_df_calo = []
10
11 # collect calorimetry data
12 for file in (path_data / "calorimetry").iterdir():
13     # info
14     print(file.name)
15     # get data
16     this_calo_data = basics.get_data(file)
17     # add file information
18     this_calo_data["experiment"] = file.stem  # filename w/o extension
19     # append data to list
20     list_df_calo.append(this_calo_data)
21
22 # compile calorimetry data
23 data_calorimetry = pd.concat(list_df_calo)
24
25 # get calorimetry raw data
26 data_vicat = pd.read_csv(
27     path_data / "vicat" / "vicat.csv",
28     ).iloc[:,1:]
29
30 # save "cleaned" data
31
32 # define path to "cleaned" data
33 path_data_cleaned = path_data.parent / "_cleaned"
34
35 # make folder if it doesn't exist
36 if not path_data_cleaned.exists():
37     # make the folder
38     path_data_cleaned.mkdir()
39
40 # write cleaned data to this folder
41 data_calorimetry.to_excel(
42     path_data_cleaned / "calorimetry.xlsx", index=False)
43 data_vicat.to_excel(path_data_cleaned / "vicat.xlsx", index=False)
```

https://doi.org/10.1515/9783111334608-036

Listing 29.2: Collecting *information* across multiple source files for further processing and visualization.

```python
import pathlib
import pandas as pd
from mycalohelpers import basics

# data base path
path_data = pathlib.Path().cwd().parent / "_data" / "_raw"

# init list of dataFrames
list_df_calo = []

# collect calorimetry data
info_application = pd.read_json(
    path_data / "application" / "application.json"
    )

# collect calorimetry information
# init list of dataFrames
list_info_calo = []

# collect calorimetry data
for file in (path_data / "calorimetry").iterdir():
    # info
    print(file.name)
    # get data
    _, this_calo_info = basics.get_peak(file, show_plot=False)
    # add file information
    this_calo_info["id"] = file.stem  # file name without extension
    # append data to list
    list_info_calo.append(this_calo_info)

# compile calo info
info_calo = pd.concat(list_info_calo)

# drop filename column
info_calo = info_calo.drop(columns=["file"])

# %% combine all parameters

parameters = pd.merge(
    left=info_application,
    right=info_calo,
    left_on="id",
    right_on="id"
    )


# define path to "cleaned" data
path_info_cleaned = path_data.parent.parent / "_parameters"

# make folder if it doesn't exist
if not path_info_cleaned.exists():
    # make the folder
    path_info_cleaned.mkdir()

# write cleaned data to this folder
parameters.to_excel(
    path_info_cleaned / "parameters.xlsx", index=False)
```

30 Simplicity of `matplotlib`

Problem

You want a simple representation of experimental data and information that allows a high degree of customization.

Context

Visualizing experimental results in a "meaningful" way is probably one of *the* tasks of a natural scientist, as a compelling set of visuals serves as a way to effectively communicate your findings. The highest standards are probably set by scientific journals, where only thoroughly validated and peer-reviewed research should be made available to a wider audience.

But is this actually the main part of scientific work? Most of the visualization and plotting is done "on the fly" immediately after an experiment is conducted to get an impression of the validity of a measurement. Did everything go as planned? Was there a problem with the machine? Were the correct settings used? These are the first questions that are regularly asked when new experimental results are received.

Once these criteria have been met, a typical next step is to compare multiple experiments. How does sample A compare to sample B? Is there a trend in the series transitioning from samples A through D? Why is there a deviation from the pattern in sample C?

You'd like to be able to perform these routine plotting tasks efficiently and with high quality without having to copy data from one file to another in order to compare results and visualize them according to your specific preferences.

Solution

Use properly sorted data and the `matplotlib` package to visualize your results as shown in code section 30.1. The basic function of `matplotlib` is `plot`. It takes the data points to plot and numerous additional keywords that allow for customization. Here we will go through all the samples, select a few of them, and customize the plots according to their identity.

Listing 30.1: Using `matplotlib` for custom visualizations.

```
1  import pathlib
2  import pandas as pd
3  import re
4  import matplotlib.pyplot as plt
5  import datetime
6
7  # get calorimetry data
```

https://doi.org/10.1515/9783111334608-037

```
 8   data = pd.read_excel(pathlib.Path().cwd().parent.parent /
 9       "_Data_and_code" / "_data" / "_cleaned" / "calorimetry.xlsx"
10       )
11
12   # %% plot
13
14   for sample_name, sample_data in data.groupby(by="experiment"):
15       # restrict plotting to some samples 4, 7, 9, 13
16       if not re.match(".*((04)|(07)|(09)|(13))$", sample_name):
17           # go to next experiment
18           continue
19
20       # define specific appearance for a "reference", here #09
21       if sample_name == "P23_X_009":
22           # set linstyle
23           linestyle="--"  # dashed
24           color="gray"
25       else:
26           # default linestyle
27           linestyle="-"  # solid
28           color=None
29
30       # info
31       print(sample_name)
32
33       # plot
34       plt.plot(
35           sample_data["time_s"]/60,  # -> [min]
36           sample_data["normalized_heat_flow_w_g"]*1000,  # -> [mW/g]
37           label=sample_name,
38           linewidth=1,
39           linestyle=linestyle,
40           color=color
41           )
42
43   # add range
44   plt.axvspan(0, 15, color="gray", alpha=0.2)
45
46   # indicative line
47   plt.axvline(20, color="red", alpha=0.2)
48   # add text
49   plt.text(21, 5, "Assumend_end_of\nequilibration",
50           size=8, verticalalignment="top")
51
52   # add legend
53   plt.legend(frameon=False)
54
55   # set limit
56   plt.xlim(left=0)
57
58   # add labels
59   plt.xlabel("Time_/_[min]")
60   plt.ylabel("Normalized_Heat_Flow_/_[mW/g]")
61
62   plt.text(
63       1, -0.25,
64       f"Timestamp:_{datetime.date.today()}",
65       transform=plt.gca().transAxes,
66       ha="right",
67       size=8,
```

```
68        color="gray",
69        style="italic"
70        )
71
72 # save figure
73 plt.savefig("plot.png", dpi=300, bbox_inches="tight")
```

The resulting plots can also be saved in various formats for documentation and/or publication (see Figure 30.1).

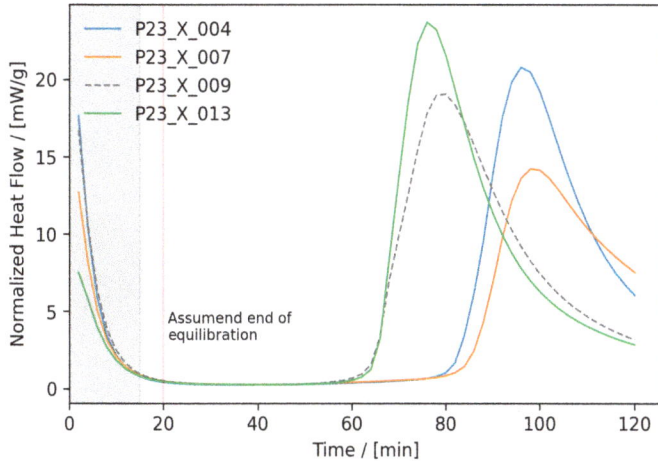

Figure 30.1: Side-by-side comparison of experimental heat flow calorimetry results including a visually distinct (dashed) guide-to-eye line.

The benefit of strong customization options comes at the cost of (some) verbosity compared to other plotting libraries.

Discussion

The flexibility and large number of options of `matplotlib` comes at the cost of complexity. Compared to other plotting libraries (see Concept 32 and Concept 33), many lines of code are needed to achieve a a visualization. From another perspective, you need this number of lines of code to get *the* visualization you have in mind as `matplotlib` is highly precise and customizable. As a natural scientist, I think it's well worth the effort lo learn it, as it allows you to generate both on-the-fly plots and publication-quality figures.

Further reading

For an overview of the possibilities that come with `matplotlib`, see their *gallery*,[1] which is a valuable resource.

---

1 https://matplotlib.org/stable/gallery/index.html

31 Creating a custom `matplotlib` style

Problem

You would like to have all your `matplotlib` plots (and built on top of the plotting libraries) styled according to defined guidelines.

Context

As outlined in Concept 30, `matplotlib` provides enormous flexibility at the cost of some verbosity. Maintaining "proper", i. e., consistent formatting of plots over a longer period of time is a challenge in itself if you do not have proper helpers.

This may not seem decisive at first, but keep in mind that these kinds of visual distractions add up. In the end, it might be much more difficult to identify crucial experimental effects if you get lost in sorting out the visual clutter.

Solution

To save you time configuring, e. g., the location of labels, default fonts and sizes, to name just a few, there's also the option of adding a custom matplotlib style file of the type *mplstyle*. Within this file, any of the parameters defined in the online template file can be configured.[1] An example of such a file is given in code section 31.1.

Listing 31.1: Example of a *mplstyle* file for configuring the appearance of `matplotlib` visualizations.

```
 1  # Plot settings
 2  axes.titlesize : 18
 3  axes.labelsize : 12
 4  lines.linewidth : 1.5
 5  lines.markersize : 10
 6  xtick.labelsize : 10
 7  ytick.labelsize : 10
 8  axes.prop_cycle : cycler(color=['c', 'm', 'y', 'k'])
 9  axes.facecolor : white
10  axes.edgecolor : gray
11  axes.labelcolor : (0.07843, 0.42745, 0.98431, 1.0)  # light blue
12  axes.spines.top : False
13  axes.spines.right : False
14  xtick.color : gray
15  ytick.color : gray
```

---

1 https://matplotlib.org/stable/tutorials/introductory/customizing.html

https://doi.org/10.1515/9783111334608-038

To make the new style called *p23* available in our plots, we need to move it to the `matplotlib`'s configuration directory and reload the style library. Once available, we can check the newly available style as described in code section 31.2.

Listing 31.2: Using matplotlib for custom visualizations.

```
1  import pathlib
2  import pandas as pd
3  import re
4  import matplotlib
5  import matplotlib.pyplot as plt
6  import datetime
7  import shutil
8
9  # %% make the stylefile available
10
11 # define path to stylefile
12 stylefile = pathlib.Path().cwd() / "p23.mplstyle"
13
14 # add user defined styles folder
15 config_path = pathlib.Path(matplotlib.get_configdir()) / "stylelib"
16 # make path if it does no exist
17 if not config_path.exists():
18     # make path
19     config_path.mkdir()
20
21 # copy user defined stylefile to matplotlib config directory
22 shutil.copy(
23         str(stylefile),
24         str(config_path / stylefile.name)
25         )
26
27 # reload the libary
28 matplotlib.style.reload_library()
29
30 # show available styles
31 for style in matplotlib.style.available:
32     # show
33     print(style)
```

Using or rather applying the newly available style *p23* is as easy as setting the `matplotlib` style according to code section 31.3.

Listing 31.3: Using matplotlib for custom visualizations.

```
1  # apply the style
2  plt.style.use(["p23"])
```

Except for the style selection and the initial effort of adding the *p23* style to the `matplotlib`'s style library, the script is unchanged compared to Concept 30. The resulting plot is shown in Figure 31.1.

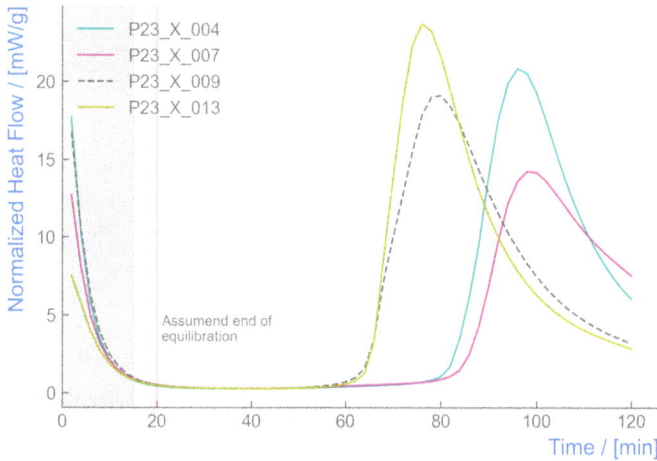

Figure 31.1: Overview of heat flow calorimetry curves using the custom *p23* matplotlib style.

Discussion

Configuring the appearance of plots using matplotlib styles is a convenient way of maintaining consistency within a project. Note that several available styles can be provided to the matplotlib.style.use function as a Python list. As a result, the keywords specified for one style are overwritten by the keywords specified for the next style. Therefore, changing the order of styles declared in the function may affect the final appearance of the plot in certain aspects.

Try adding the *bmh* style as an additional source of formatting defaults. Change the order of the styles provided and check the available options via the matplotlib.style.available attribute.

Further reading

An overview of the options available with matplotlib is given in its *gallery*,[2] which is a valuable resource.

The SciencePlots-package[3] configures matplotlib to generate plots according to the requirements of major scientific journals.

---

2 https://matplotlib.org/stable/gallery/index.html
3 https://pypi.org/project/SciencePlots/

32 Convenience of seaborn

Problem

You want a quick representation of experimental data and information and are willing to sacrifice some options for customization.

Context

Often, a comparison of experimental data is required as a first step for further evaluation. Therefore, you might want to compare the most recent set of experimental data with all data originating from the same type of experimental procedure, or you may want to limit the comparison to a selection. In either of these scenarios, the ability to perform basic plotting tasks will save you a considerable amount of time.

Solution

Use properly sorted data and seaborn to visualize your results, as shown in code section 32.1. The seaborn package works best, i. e., according to the expectations and needs of us natural scientists, when relying on the *long-data format*.[1] In this file, all tabular data is formatted to contain all numeric values in one column and one or more columns representing the context of the values. In our example, the normalized heat flow data column represents the value column, and the experiment column defines the context of those values.

Listing 32.1: Using the seaborn package for visualizations.

```
1  import pathlib
2  import pandas as pd
3  import re
4  import matplotlib.pyplot as plt
5  import datetime
6  import seaborn as sns
7
8  # mpl style also influences seaborn plot
9  plt.style.use(["p23"])
10
11 # get calorimetry data
12 data = pd.read_excel(pathlib.Path().cwd().parent.parent /
13     "_Data_and_code" / "_data" / "_cleaned" / "calorimetry.xlsx"
14     )
15
16 # scale data
```

---

1 https://en.wikipedia.org/wiki/Wide_and_narrow_data

https://doi.org/10.1515/9783111334608-039

```
17  data["Time_/_[min]"] = data["time_s"]/60
18  data["Normalized_Heat_Flow_/_[W/g]"] =\
19      data["normalized_heat_flow_w_g"]*1000
20
21  # plot
22  sns.lineplot(
23      data=data[data["experiment"].\
24          str.match(".*((04)|(07)|(09)|(13))$")],
25      x="Time_/_[min]",
26      y="Normalized_Heat_Flow_/_[W/g]",
27      hue="experiment"
28      )
29
30  # add range
31  plt.axvspan(0, 15, color="gray", alpha=0.2)
32
33  # indicative line
34  plt.axvline(20, color="red", alpha=0.2)
35  # add text
36  plt.text(21, 5, "Assumend_end_of\nequilibration", size=8,
37          verticalalignment="top")
38
39  # add legend
40  plt.legend(frameon=False, title="Experiment_ID",
41          title_fontsize="small", alignment="left")
42
43  # set limit
44  plt.xlim(left=0)
45
46  # add text
47  plt.text(
48      1, -0.25,
49      f"Timestamp:_{datetime.date.today()}",
50      transform=plt.gca().transAxes,
51      ha="right",
52      size=8,
53      color="gray",
54      style="italic"
55      )
56
57  # save figure
58  plt.savefig("plot.png", dpi=300, bbox_inches="tight")
```

The resulting plot is shown in Figure 32.1.

Discussion

The numerous out-of-the-box features of seaborn come at the cost of a reduced customization experience. However, compared to other plotting libraries (see Concept 30
and Concept 33), only a few lines of code are needed to achieve an impressive visualization. This is true if the data to be displayed is structured in the required "long" format
and your target plot is close to the built-in available plot types. To account for the lack of
customization experienced, seaborn plots can be further modified using matplotlib.

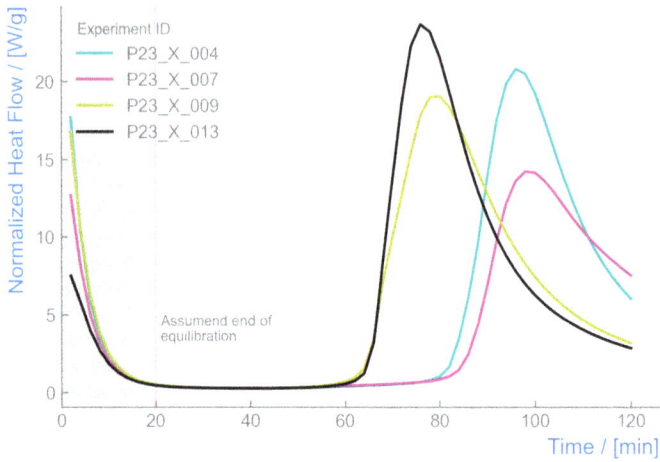

Figure 32.1: Visualization of heat flow calorimetry curves using the `seaborn` package.

Also, the previously defined `matplotlib` style can be used to customize the appearance of seaborn plots.

Further reading

For an overview of the possibilities that come with seaborn, see its *gallery*,[2] which is a valuable resource.

---

2 https://seaborn.pydata.org/examples/index.html

33 Interactivity of `plotly`

Problem

You want to have an interactive and fast representation of experimental data and information and are willing to sacrifice some options for customization.

Context

Often, a comparison of experimental data is required as a first step towards further evaluation. For example, you may want to compare the most recent set of experimental obtained with all the data from the same type of experimental procedure, or you may want to limit the comparison to a selection. In either of these scenarios, the ability to perform basic plotting tasks will save you a considerable amount of time.

Solution

Use properly sorted data and the `plotly`-package[1] to interactively visualize your results as shown in code section 33.1.

Listing 33.1: Using `plotly` for visualizations.

```
1  import pathlib
2  import pandas as pd
3  import plotly.express as px
4
5  # get calorimetry data
6  data = pd.read_excel(pathlib.Path().cwd().parent.parent /
7      "_Data_and_code" / "_data" / "_cleaned" / "calorimetry.xlsx"
8      )
9
10 # scale data / unit conversion
11 data["Time_/_[min]"] = data["time_s"]/60
12 data["Normalized_Heat_Flow_/_[W/g]"] =\
13     data["normalized_heat_flow_w_g"]*1000
14
15 # make lineplot
16 fig = px.line(
17     data[data["experiment"].str.match(".*((04)|(07)|(09)|(13))$")],
18     x="Time_/_[min]",
19     y="Normalized_Heat_Flow_/_[W/g]",
20     color="experiment",
21     title="Overview_line_plot",
22     markers=True,
23     symbol="experiment",   # symbols by experiment
```

---

1 https://pypi.org/project/plotly/

https://doi.org/10.1515/9783111334608-040

```
24      )
25
26  # save figure as interactive html
27  fig.write_html("plot.html")
28
29  # save figure as png (requires the "kaleido"-package)
30  fig.write_image("plot.png")
```

The real benefit of working with `plotly` or its lower-level interface `plotly-express` is the ability to create interactive plots in only a few lines of code. To maintain the inter-activity suggested in Figure 33.1, where the individual viewer can define which experiments and which areas are examined, a `plotly` figure can be exported to hypertext markup language (HTML) format, which can be opened by browser applications such as Google Chrome, Microsoft Edge, and the like. In the example, deselected experiments are displayed in the legend by a grayed out appearance. In addition, hovering the mouse pointer over the individual plot lines will highlight the characteristics of the corresponding data point in an info box. There are numerous options for configuring both the level of interactivity and the information displayed as hover texts.

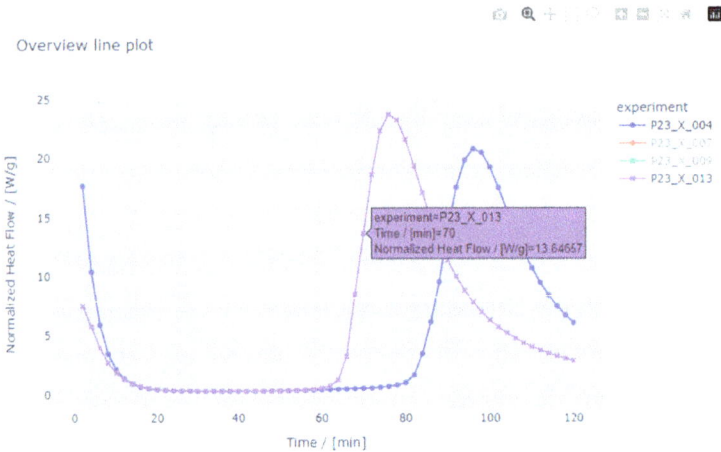

Figure 33.1: Interactive image export of a `plotly` figure as a HTML-file. Hovering over the lines provides additional information about the measured point. Clicking on the legend entries on the right-hand side allows you to show or hide curves related to an individual sample.

Additionally, `seaborn` allows you to save the plots as static images, as shown in Figure 33.2.

Overview line plot

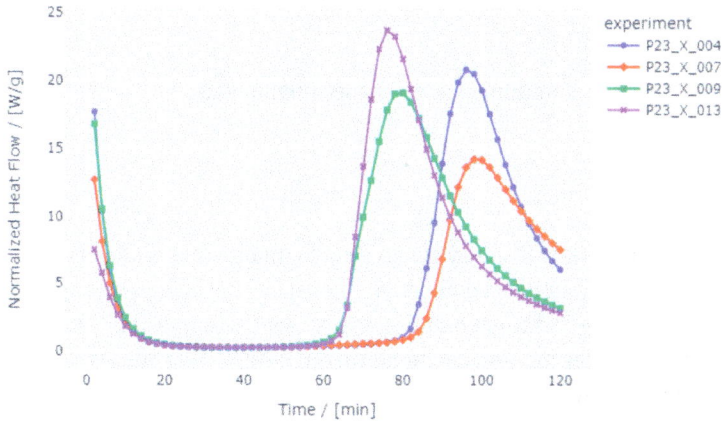

Figure 33.2: Static image export of a `plotly` figure.

Discussion

The numerous out-of-the-box features of `plotly` come at the cost of a reduced customization experience. However, compared to other plotting libraries (see Concept 30 and Concept 32), only a few lines of code are needed to achieve an impressive interactive visualization. This holds true if the data you want to display is structured in the required format and your target plot is close to the built-in available plot types.

Further reading

For an overview of the options available with `plotly`, see its *gallery*,[2] which is a valuable resource.

---

2 https://plotly.com/python/

34 Representing multidimensional data

Problem

You want to visually represent multidimensional experimental data.

Context

Presenting multidimensional data in a way that is easy to understand is one of the key challenges in the field of formulation chemistry. At one end of the data to be displayed is the chosen composition of a certain mixture. In the present example, this mixture is characterized according to a certain set of methods (in our case, a viscosity, a maximum pull-off force and the heat flow calorimetry characteristics of the peak position and the corresponding normalized heat flow are determined) as shown in Figure 34.1.

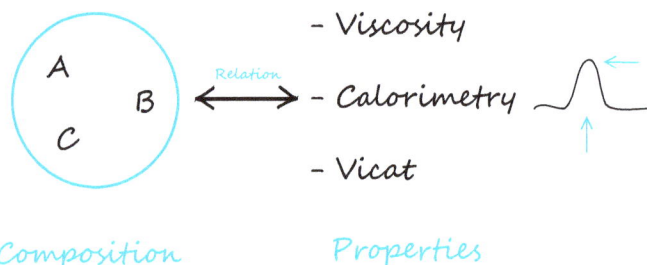

Figure 34.1: Characteristics of an individual formulation. Compositions can be represented in terms of relative content. Measured properties/characteristics are assigned to these specific components A, B and C.

Here, for the sake of argument, we have three initially determined values (also referred to as inputs) and four measured or resulting parameters (also referred to as outputs). In a more realistic setting, the number of independently defined inputs and resulting outputs may be much larger. In addition to these numerically available numbers, you may have certain hypotheses about *cause and effect* in mind. For example, the ratio of two components might be correlated with an experimentally determined characteristic.

Solution

A convenient way to visualize multidimensional data is provided by the plotly-package's[1] parallel_coordinates function to interactively (if saved as HTML) visu-

---

1 https://pypi.org/project/plotly/

https://doi.org/10.1515/9783111334608-041

alize your results as shown in 34.1. Derived characteristics, such as concentration ratios of the components, can be added as additional columns to the pd.DataFrame.

Listing 34.1: Visualizing multidimensional data using `parallel_coordinates`.

```
1  import pathlib
2  import numpy as np
3  import pandas as pd
4  import matplotlib.pyplot as plt
5  import plotly.express as px
6  import seaborn as sns
7
8  #  get composition
9  composition = pd.read_excel(pathlib.Path().cwd().parent.parent /
10     "_Data_and_code" / "_data" / "_raw" / "composition" /\
11     "composition.xlsx",
12     usecols="B:E"
13     )
14
15 # get parameters
16 parameters = pd.read_excel(pathlib.Path().cwd().parent.parent /
17     "_Data_and_code" / "_parameters" / "parameters.xlsx"
18     )
19
20 # combine composition and parameters to "data"
21 data = pd.merge(composition, parameters)
22
23 # show columns
24 for c in data.columns:
25     print(c)
26
27 # add more columns for hypothesis testing
28 data["A:B"] = data["A"]/data["B"]
29 data["A:C"] = data["A"]/data["C"]
30 data["B:C"] = data["B"]/data["C"]
31
32 # configure and save plot
33 # parallel coordinates plot
34 fig = px.parallel_coordinates(
35     data,
36     color="normalized_heat_flow_w_g",  # color by heat flow
37     dimensions=[ # select columns (order left to right maintained)
38         "A",
39         "C",
40         "B",
41         "A:B",
42         "A:C",
43         "B:C",
44         "time_s",
45         "viscosity_mpas",
46         "pull_off_force_N",
47         ],
48     labels={
49         "time_s": "Calorimetry_Peak\nPosition_/_[s]",
50         },
51     color_continuous_scale=px.colors.sequential.Bluered,
52     range_color=(
53         data["normalized_heat_flow_w_g"].min(),
```

```
54          data["normalized_heat_flow_w_g"].max()
55        ),
56      # https://plotly.com/python/builtin-colorscales/
57      color_continuous_midpoint=data["normalized_heat_flow_w_g"].\
58          median()
59    )
60
61  # save as html file
62  fig.write_html("plot.html")
63
64
65  # findings
66  # playing in html: correlation "pull_off_force_N" <-> "B:C"
67  # add scatter plot
68  sns.scatterplot(
69      data=data,
70      x="B:C",
71      y="pull_off_force_N",
72      hue="normalized_heat_flow_w_g"
73    )
74
75  # save plot
76  plt.savefig("plot_scatter.png", dpi=300, bbox_inches="tight")
77
78  # %% correlaction matrix
79
80  corr = data.corr(numeric_only=True)
81
82  # Getting the Upper Triangle of the co-relation matrix
83
84
85  plt.figure(figsize=(8,6))
86  sns.heatmap(corr,
87      # https://seaborn.pydata.org/tutorial/color_palettes.html
88      cmap=sns.color_palette("coolwarm", as_cmap=True),
89      vmin=-1.0, vmax=1.0,
90      annot=True,
91      annot_kws={"fontsize" : 8},
92      fmt=".2f",
93      mask=np.triu(corr),
94      square=True)
95
96  plt.gca().set_facecolor("white")
97  plt.grid(None)
98
99  # save plot
100 plt.savefig("plot_matrix.png", dpi=300, bbox_inches="tight")
101
102 # filter correlations
103 kk = corr.filter(regex="[ABC]$", axis=0)\
104     .filter(regex="^((?![ABC]).)*$")\
105     .reset_index()\
106     .melt(id_vars=["index"])\
107     .sort_values(by="value", ascending=False)\
108     .reset_index(drop=True)
109
110 print(kk)
111
112 #      index            variable      value
113 # 0    B:C        pull_off_force_N  0.545471
```

```
114  # 1        B          pull_off_force_N   0.468142
115  # 2    A:C                viscosity_mpas   0.267888
116  # 3    B:C  normalized_heat_flow_w_g   0.229945
```

In the resulting parallel coordinates plot shown in Figure 34.2, the numerical values are
displayed vertically in one column each. Results corresponding to one *observation*, i. e.,
one row in the data table, i. e., pd.DataFrame are represented by an individual line. The
number of rows is equal to the number of experiments considered. Using the parallel coordinates plot, it seems possible to summarize the key findings of an entire formulation
project (with limited scope, of course). In this plot, "parallel, i. e.,non-intersecting" lines
between neighboring colors indicate a high correlation coefficient. Conversely, "crossing" lines indicate a negative correlation.

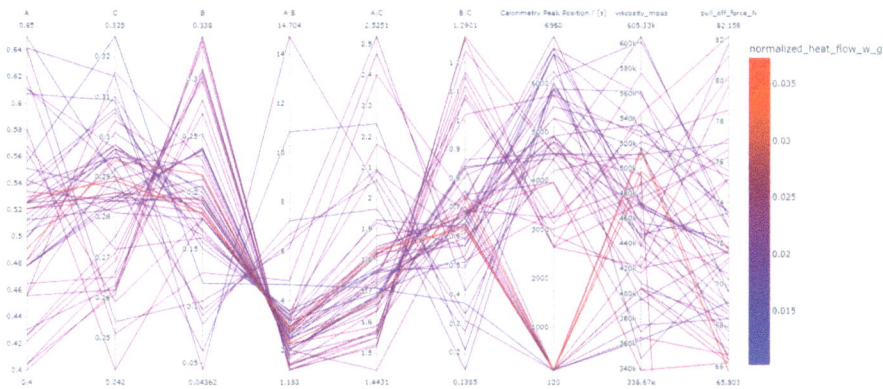

Figure 34.2: Display of multidimensional data related to the compositions and experimental properties of
the formulations studied using the plotly's parallel_coordinates function.

These correlations can also be verified using a conventional correlation matrix as
accessible from seaborn's heatmap function. To familiarize yourself with the presentation of high and low correlation coefficient values, the low correlation value between
the variables A and B:C is shown next to the high correlation value between the variables B:C and B in Figure 34.3 in the parallel coordinates plot and in Figure 34.4 in the
correlation heatmap for comparison.

Discussion

Once again, the interactivity of plotly plots shines in the parallel coordinate representation of multidimensional data. It allows a true "hands-on" experience for understanding
your project data. Key features include

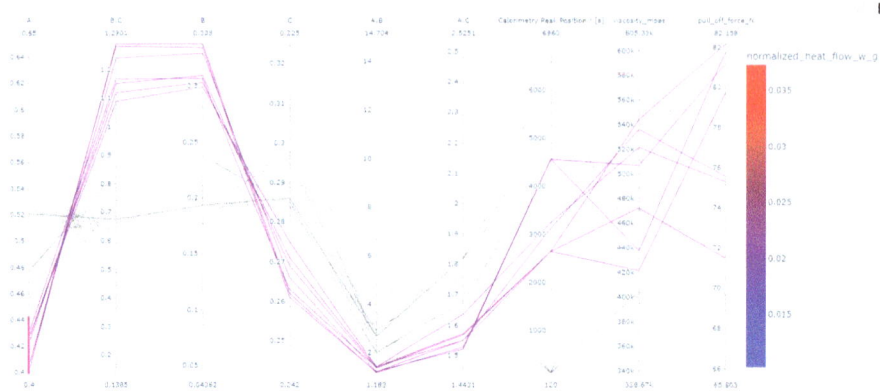

Figure 34.3: Same as Figure 34.2 but with dimensions rearranged for highlighting correlations.

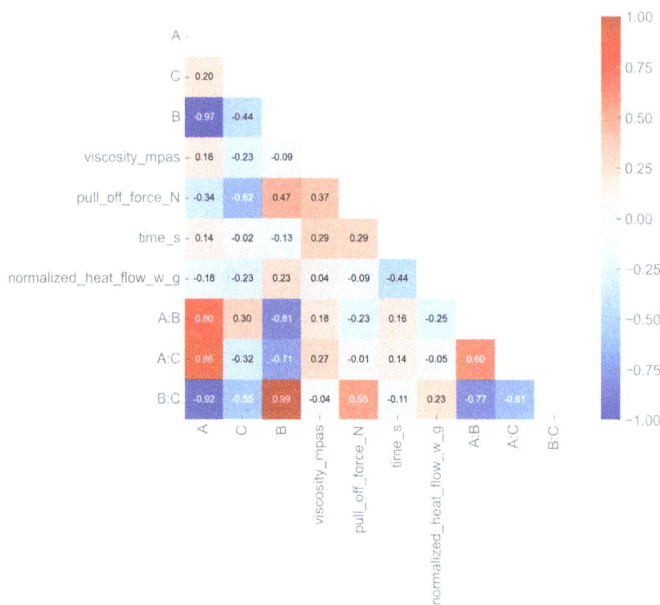

Figure 34.4: Heatmap of correlation coefficients within the example dataset representing the relationships between the ternary formulation consisting of components A, B and C, and the measured experimental characteristics.

- selecting a certain range of values for one or more target characteristics in the sense of a "target corridor" by left-clicking the mouse. Only observations, i. e., experiments falling within this range will be highlighted, while the remaining experiments will be grayed out (see Figure 34.3).
- shifting the above mentioned target ranges up and down along one of the axes. This allows answering questions of the following type:

- How does a particular output characteristic change when the concentration of a certain input is increased (or decreased)?
- If a particular output characteristic is to be modified in a certain direction, how must the inputs be changed?
- shifting, i. e., rearranging column orders by dragging and dropping to reveal either a positive correlation (parallel lines) or an anticorrelation (crossing lines).

Further reading

For an overview of other options available with `plotly`, see the project's *gallery.*[2]

---

2 https://plotly.com/python/

35 Representing multidimensional data in a funny way

Problem

You want to visually represent multidimensional experimental data in a way that is understandable, preferably to a non-expert audience.

Context

Presenting multidimensional data in an easily understandable way to an audience unfamiliar with the details of your data, e. g., in the context of a broader conference talk is one of the key challenges in the field of scientific data presentation. The high dimensionality of the data presented is also likely to cause a loss of your audience's attention. Since oversimplifying, i. e., reducing the dimensionality may not be feasible or your preferred solution, the use of a *Chernoff* face type representation may be worth a try.

Solution

Use the `ChernoffFace` package[1] to generate Chernoff face type diagrams. The key idea here is to display multivariate data in the form of a human-like cartoon face. Individual features such as eyes, ears, mouth, and nose represent values of the variables by their shape, size, placement, and orientation. The idea behind using faces is that humans can easily recognize faces and notice small changes.

All that remains to be done is to scale your experimental data properly so that it can be displayed using the Chernoff faces as shown in code section 35.1.

Listing 35.1: Using the `ChernoffFace` package to visualize multidimensional data.

```
import pathlib
import numpy as np
import pandas as pd
import matplotlib.pyplot as plt
import matplotlib.cm
from sklearn import preprocessing

# import
from ChernoffFace import chernoff_face, single_chernoff_face


# %% get calorimetry data
```

---

1 https://pypi.org/project/ChernoffFace/

https://doi.org/10.1515/9783111334608-042

```
13
14   #  get composition
15   composition = pd.read_excel(pathlib.Path().cwd().parent.parent /
16       "_Data_and_code" / "_data" / "_raw" / "composition" /\
17       "composition.xlsx",
18       usecols="B:E"
19       )
20
21   # get parameters
22   parameters = pd.read_excel(pathlib.Path().cwd().parent.parent /
23       "_Data_and_code" / "_parameters" / "parameters.xlsx"
24       )
25
26   # combine composition and parameters to "data"
27   data = pd.merge(composition, parameters)
28
29   # filter for complete datasets
30   data = data.dropna(axis=0)
31
32   data = data.set_index("id")
33
34   # normalize columns --> scale results from 0.1 to 0.9 in each colum
35   dev=0.1
36   # init "scaler"
37   min_max_scaler = preprocessing.MinMaxScaler(
38       feature_range=(dev, 1-dev))
39   # scale columns
40   for c in data.columns:
41       # Min Max Scaling
42       data[c] = min_max_scaler.fit_transform(
43           np.array(data[c]).reshape(-1,1))
44
45   selected = data.head(12)
46
47   # Make Chernoff faces (all)
48   fig = chernoff_face(
49       data=selected,
50       n_columns=4,
51       long_face=False,
52       titles=[str(i) for i in selected.index],
53       color_mapper=matplotlib.cm.tab20b,
54       figsize=(8, 5),
55       dpi=200
56       )
57
58   plt.text(0, 1.5, "Composition_and_Results", fontweight="bold",
59             transform=fig.axes[0].transAxes, size=16)
60
61   # tune layout and show
62   fig.tight_layout()
63   plt.show()
64
65   # Make Chernoff faces (composition)
66   fig = chernoff_face(
67       data=selected.filter(regex="[ABC]"),
68       n_columns=4,
69       long_face=False,
70       titles=[str(i) for i in selected.index],
71       color_mapper=matplotlib.cm.tab20b,
72       figsize=(8, 5),
```

```
73        dpi=200
74        )
75
76   plt.text(0, 1.5, "Compositions", fontweight="bold",
77            transform =fig.axes[0].transAxes, size=16)
78
79   # show
80   plt.show()
81
82
83   # %% results
84
85   # Make Chernoff faces (composition)
86   fig = chernoff_face(
87        data=selected.filter(regex="^((?![ABC]).)*$"),
88        n_columns=4,
89        long_face=False,
90        titles=[str(i) for i in selected.index],
91        color_mapper=matplotlib.cm.tab20b,
92        figsize=(8, 5),
93        dpi=200
94        )
95
96   plt.text(0, 1.5, "Results", fontweight="bold",
97            transform=fig.axes[0].transAxes, size=16)
98
99
100  # Display
101  fig.tight_layout()
102
103  plt.show()
104
105  # validate similarity of experiments 3 and 9 via table
106  print(selected.filter(regex="00(1|3|4)", axis=0).T)
107
108
109  # %%
110
111  # "FaceSize": 0.5,
112  # "ForeheadShape": 0.5,
113  # "EyesVerticalPosition": 0.5,
114  # "EyeSize": 0.5,
115  # "EyeSlant": 0.5,
116  # "LeftEyebrowSlant": 0.5,
117  # "LeftIris": 0.5,
118  # "NoseLength": 0.5,
119  # "MouthSmile": 0.5,
120  # "LeftEyebrowTrim": 0.5,
121  # "LeftEyebrowRaising": 0.5,
122  # "MouthTwist": 0.5,
123  # "MouthWidth": 0.5,
124  # "RightEyebrowTrim": 0.5,
125  # "RightEyebrowRaising": 0.5,
126  # "RightEyebrowSlant": 0.5
127  # "RightIris": 0.5,
128
129
130  for i, val in enumerate([.2, .5, .8]):
131      # init figure
132      fig = plt.figure(figsize=(2, 3), dpi=200)
```

```
133    # plot
134    single_chernoff_face({
135            "FaceSize" : val,
136            "EyeSize" : val,
137            "MouthWidth" : val,
138            "MouthSmile" : val,
139            "RightIris" : 0.25,  # look "right" on right eye
140            "LeftIris" : 0.25  # look "right" on left eye
141            },
142        rescale_values=False,
143        # color_mapper=matplotlib.cm.tab20b,
144        figure=fig
145        )
146    # title
147    plt.title(f"All_values_{val}.")
148    # save
149    plt.savefig(f"faces/all_{val}.png", dpi=300, bbox_inches="tight")
150
151
152  # loop experiments
153  for i, (experiment_no, experiment) in enumerate(
154        selected.filter(regex="^((?![ABC]).)*$").iterrows()):
155
156    # init figure
157    fig = plt.figure(figsize=(2, 2), dpi=200)
158    # plot
159    face = single_chernoff_face({
160            "FaceSize" : experiment["viscosity_mpas"],
161            "ForeheadShape" : experiment["pull_off_force_N"],
162            "MouthWidth" : experiment["time_s"],
163            "MouthSmile" : experiment["normalized_heat_flow_w_g"],
164            },
165        rescale_values=False,
166        color_mapper=matplotlib.cm.tab20b,
167        figure=fig
168        )
169    # add title
170    plt.title(experiment_no)
171    # save
172    plt.savefig(f"faces/{experiment_no}.png",
173                  dpi=300, bbox_inches="tight")
```

To get an idea of the implications of the numerical values, the faces of the parameters *FaceSize, EyeSize, MouthWidth* and *MouthSmile* for values of 0.2, 0.5 and 0.8 are shown in Figure 35.1.

Similarly, more or less similar formulations, i. e., compositions of ternary compounds consisting of A, B and C can be readily visually identified from the Figure 35.2.

Get the numeric values represented in Figure 35.2 from the above code section 35.1 and confirm the "visually experienced similarity" also in the underlying pd.DataFrame.

All values 0.2.   All values 0.5.   All values 0.8.

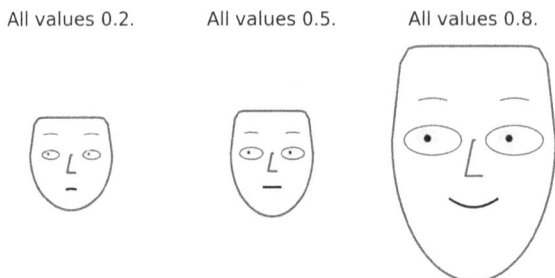

Figure 35.1: Increasing the *FaceSize*, *EyeSize*, *MouthWidth* and *MouthSmile* properties of *Chernoff* faces in three steps of 0.2, 0.5 and 0.8.

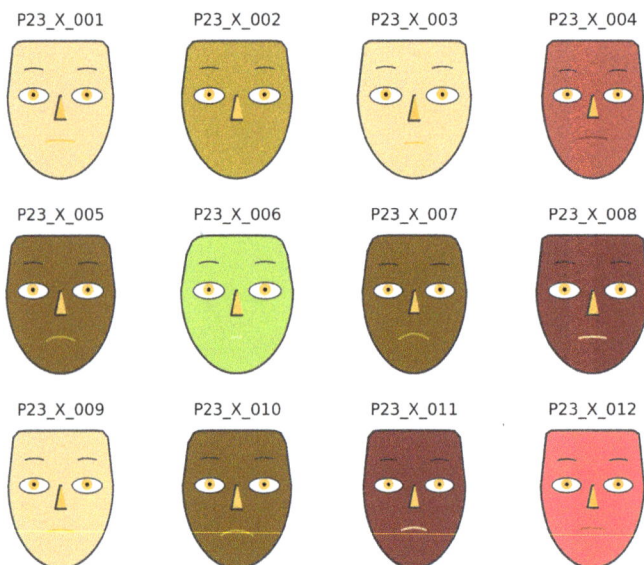

Figure 35.2: Presentation of the experimental properties of the studied formulations represented by the `ChernoffFace` package.

Discussion

At the time of writing, the use of Chernoff faces to represent multidimensional data is not a standard in the natural sciences, in particular chemistry and materials science. However, due to the advantages described above, it is expected that they could be a valuable technique, especially in the context of conveying complex experimental results within a short time frame to an audience (also referred to as *Storytelling with data*[2]) not familiar with "your" topic in full detail.

---

2 There is a whole set of literature related to this topic.

Further reading

Several applications of the method of graphical representation of multivariate data via Chernoff faces are described in scientific literature.[3]

---

3 https://www.sciencedirect.com/science/article/abs/pii/B9780127347509500137

Approaching the scientific questions (modeling and recommendation)

So far, we have dealt mostly with "organizational" issues and laying the foundations for gaining value from thoroughly planned, executed, and analyzed experiments. The following concepts will foucus on modeling aspects. The goal of these models is to enable you to make informed predictions about the (most likely) expected behavior of your system, taking into account *all* experimental evidence.

To be very clear, the predictions, i. e., expected values are not "out of the blue", but based on *your* experimental data and the results of your analysis as indicated in Figure 5. This also emphasizes the importance of running experiments thoughtfully and accurately – every single experimental data point matters!

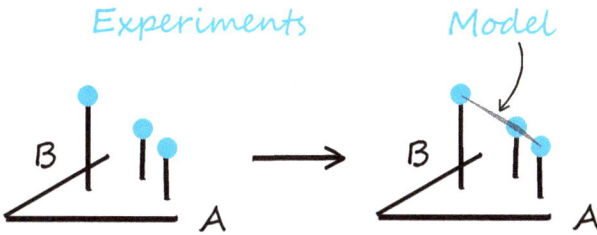

Figure 5: From individual experiments to a continuous model.

On the other hand, since potentially the entire set of experimental results is considered in the modeling, the tendency to over-interpret results related to one experimental composition can be reduced. If only one of the results does not quite "fit in", it is recommended to repeat the experiment.

https://doi.org/10.1515/9783111334608-043

36 Picking relevant data and information

Problem

To prepare for the modeling steps introduced in the following, you want to combine all input and output information into a single table.

Context

To build a model of your studied system, you need to have "all" information at your fingertips in order to mathematically describe the relationships between *inputs* and *outputs*. In other words: You would like to understand how your system responds, i. e., the output when you change the composition, i. e., the input.

At some point, you're going to have to draw a line in the sand. What do you consider "relevant" in this context, given your domain knowledge and experience in this area? There are also commercial aspects and business decisions to consider, at least if you are active in industrial research. Whatever the inputs and outputs, it is recommended to work your way from simpler to more complex models.

On the input side, we want to focus on the composition in terms of the relative contents of A, B and C first. If this does not explain the observed effects, the inputs could be extended to include characteristics of the mixing procedure or other processing parameters introduced during formulation development. If this still does not explain the observed effects, the inputs could be extended to include the batches and the corresponding ages of the components A, B and C.

On the output side, it is good practice to limit the output to characteristics that are relevant to specific scientific or business related objectives. For example, you may want to exceed a certain material pull-off strength value to label your adhesive as "extra strong". At the same time, it should be as inexpensive as possible and allow for a certain processing time. The processing time requirement can be translated – in terms of measurement – into a minimum value of Vicat time and peak position derived from heat flow calorimetry.

Solution

In our showcase example, the combination of the prepared inputs and outputs is achieved using the pandas package as shown in code section 36.1. With these in hand, creating visualizations is a straightforward task (see Concept 30, Concept 32 and Concept 33).

https://doi.org/10.1515/9783111334608-044

Listing 36.1: Picking relevant *data* and *information* using pandas.

```python
import pathlib
import pandas as pd
import plotly.express as px


# %% collect data from individual sources

# define base path to data
data_base_path = pathlib.Path().cwd().parent.parent\
    / "_Data_and_code"

#  get composition
composition = pd.read_excel(
    data_base_path / "_data" / "_raw" / "composition"\
    / "composition.xlsx",
    usecols="B:E"
    )

# get parameters (except Vicat)
parameters_i = pd.read_excel(pathlib.Path().cwd().parent.parent /
    "_Data_and_code" / "_parameters" / "parameters.xlsx"
    )

# get parameters (only Vicat)
parameters_ii = pd.read_excel(pathlib.Path().cwd().parent.parent /
    "_Data_and_code" / "_parameters" / "vicat.xlsx",
    names=["id", "vicat_set_min"]  # set column names explicitly
    )


# %% combine composition and parameters to "data"

merged = pd.merge(composition, parameters_i)
merged = pd.merge(merged, parameters_ii)

# copy fo later use
data = merged.copy()

# optionally: filter for complete datasets
data = data.dropna(axis=0)

# set "id" column as index
data = data.set_index("id")

# save
data.to_excel(data_base_path / "_data" / "_cleaned"\
              / "data_and_parameters.xlsx",)

# restrict to "first" experiments (--> mock eraly project phase)
data = data.filter(regex=".*[01]\d$", axis=0)


# %% parallel coordinates plot

# plot
fig = px.parallel_coordinates(
    data_frame=data,
```

```
58         color="time_s"
59        )
60
61 # write html file
62 fig.write_html("first_experiments.html")
63
64
65 # %% ternary scatter
66
67 # plot
68 fig = px.scatter_ternary(
69        data,
70        a="A",
71        b="B",
72        c="C",
73        hover_name=data.index,
74        color="time_s",
75        size="time_s",
76        size_max=15,
77        title="time_s"
78        )
79
80 # write html file
81 fig.write_html("first_experiments_ternary.html")
82
83
84 # %% same info from vicat and calorimetry
85
86 import matplotlib.pyplot as plt
87 import seaborn as sns
88
89 # use custom style
90 plt.style.use("p23")
91
92 # get experiment run
93 merged["run_#"] = merged["id"].str.extract("(\d+)$").astype(int)
94
95 # assign class
96 def assign_experiment_set_number(experiment_run : int):
97        # assign
98        if experiment_run < 10:
99            return "Set_1"
100       elif experiment_run < 25:
101           return "Set_2"
102       else:
103           return "Set_3"
104
105 # get assignment
106 merged["set_#"] = merged["run_#"].\
107       apply(assign_experiment_set_number)
108
109 # initialize plot
110 fig, axs = plt.subplots(1, 3,
111               sharey=True, sharex=True, figsize=(8,3))
112
113 # define x and y variables
114 x = 'time_s'
115 y = 'vicat_set_min'
116
117 # add scatterplots
```

```
118   sns.scatterplot(
119       data=merged[merged["set_#"].str.contains("1")],
120       x=x, y=y,
121       alpha=0.5,
122       ax=axs[0]
123       )
124
125   sns.scatterplot(
126       data=merged[merged["set_#"].str.contains("1|2")],
127       x=x, y=y,
128       alpha=0.5,
129       hue="set_#",
130       ax=axs[1]
131       )
132
133   sns.scatterplot(
134       data=merged,
135       x=x, y=y,
136       alpha=0.5,
137       hue="set_#",
138       ax=axs[2]
139       )
140
141   # Annotation
142   # https://matplotlib.org/stable/tutorials/text/annotations.html
143   axs[0].text(
144       6000, 80,
145       "_Workability_+_",
146       ha="left",
147       va="bottom", rotation=70, size=8,
148       bbox=dict(
149           boxstyle="rarrow,pad=0.5",
150           fc="lightblue", ec="gray",
151           lw=1
152           ))
153
154   axs[1].get_legend().remove()
155   axs[1].set_ylabel("")
156   axs[0].set_xlabel("")
157   axs[1].set_xlabel("")
158
159   # Put a legend to the right of the current axis
160   axs[2].legend(
161       loc='center_left',
162       bbox_to_anchor=(1, 0.5),
163       title="Experimental_set_#",
164       alignment="left",
165       frameon=False
166       )
167
168
169   plt.tight_layout()
170
171   # save
172   plt.savefig("vicat_vs_calorimetry.png",
173               dpi=300, bbox_inches="tight")
```

An alternative to manually assembling inputs and outputs using the above procedure is to store both inputs and outputs in a dedicated Relational Database Management System (RDBMS) or other solutions tailored to your needs.

Discussion

It is critical to recognize, that requirements may change during the life of the project. To stick with the example above, a rather vague requirement such as "workability" might have to be increased above the level initially assumed based on feedback received during the project. In terms of calorimetry and vicat characteristics, this would result in higher values for both properties.

Also, having both inputs and outputs together allows for an easy monitoring or project progress as indicated in Figure 36.1. All too often, most new insight is gained from an initial set (here "Set 1") of experiments. Follow-up and additional experiments may confirm the findings from the initial experiments and occasionally extend the accessible output ranges up and down to some extent.

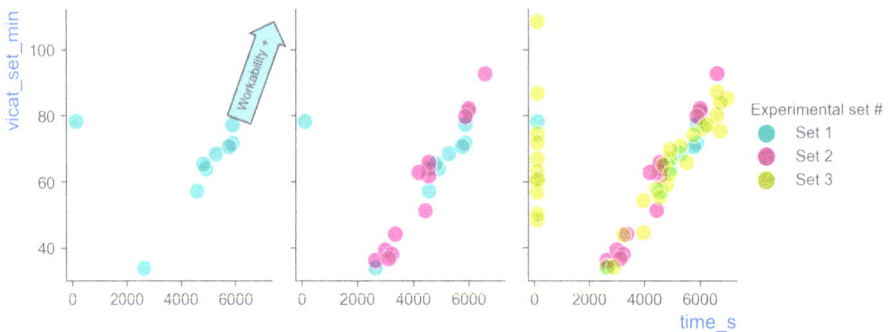

Figure 36.1: Experimental characteristics of a project summarized. The "Sets" referenced therein refer to groups of experiments run at different stages of a project, such as "identifying trends / screening", "restricting to focus area" and "fine-tuning".

If you are experiencing this "measuring within known limits" scenario, it is time to reconsider your entire experimental setup. Among the questions you might want to ask yourself are the following:
- Does the formulation have to be extended to include another component D to move into a new output range?
- Is the available experimental equipment precise enough to resolve the changes introduced by the variation of inputs?

Further reading

Typically, not all the features present in your dataset are important to get the best model performance. In addition to relying on domain and expert knowledge about which characteristics are relevant in a given dataset, there are the more systematic approaches of *feature selection*[1] and *feature engineering*.[2]

The systematic, however, generally works best for larger datasets. In the natural sciences, we are all too often *not* in the convenient position of having hundreds, thousands or even more observations, i. e., experiments available. In most cases, we have to settle for much less, as creating additional observations means using expensive materials and labor in the lab. Therefore, a combined approach of both domain knowledge and systematic methods is proposed.

---

1 https://pypi.org/project/featurewiz
2 https://pypi.org/project/auto-feature-engineering/

37 Building a model with `gplearn`

Problem

You want to build a model "connecting" the inputs and – considered relevant – outputs based on experimental data to improve your understanding of the system being analyzed.

Context

Let's assume, we've run a decent number of experiments, i. e., varied the relative contents on the *input* end of our exemplary formulation and collected all the – considered – relevant characteristics on the *output* end of things. At this point, it is time to establish the "relation" between inputs and outputs. In natural scientific terms, this is also associated with expressions such as
- determining structure-property relationships,
- mapping cause to effect,
- finding out the *Why*

and others. The traditional way of establishing this connection between inputs and outputs is based on domain, i. e., expert knowledge, which suggests that, e. g., the ratio or difference of two components on the input side is related to some characteristic on the output side. The approach outlined below describes how this hypothesis can be made in a systematic way and checked against the existing experimental data (see Figure 37.1).

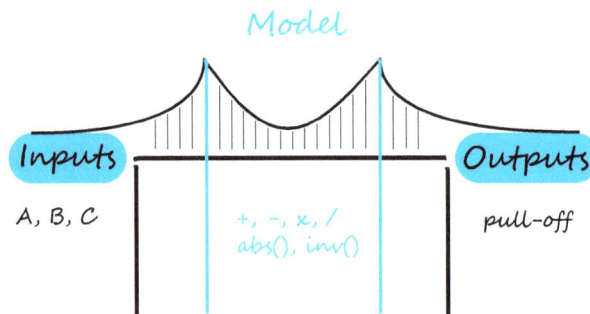

Figure 37.1: Using a model to connect the *inputs* and *outputs* of a system under study.

https://doi.org/10.1515/9783111334608-045

Solution

Use the gplearn package[1] to implement genetic programming in Python. In particular, it focuses on solving symbolic regression problems. The latter is a machine learning technique that aims to identify an underlying mathematical expression that best describes a relationship.

The key idea is as easy as selecting the basic inputs and *the* output. In our case, we select the relative contents of components A, B and C as the inputs and the measured pull-off force as the output.[2] In addition to inputs and outputs, we need to define (or allow) a certain set of functions according to which the inputs can be combined. As shown in code section 37.1, we allow addition, subtraction, multiplication and division, among others, to generate new features based on the relative contents of A, B, and C. Given these settings, gplearn begins by building a population of naive random formulas to represent a relationship between the known independent variables, i. e., inputs and their dependent variable targets, i. e., outputs in order to predict new data.

Listing 37.1: Using gplearn to model relationships between inputs and measured experimental outputs.

```
1  import pathlib
2  import numpy as np
3  import pandas as pd
4  import seaborn as sns
5  import matplotlib.pyplot as plt
6
7
8  # %% collect data from individual sources
9
10 # define base path to data
11 data_base_path = pathlib.Path().cwd().parent.parent\
12     / "_Data_and_code"
13
14 # get data
15 data = pd.read_excel(data_base_path / "_data" / "_cleaned"\
16     / "data_and_parameters.xlsx")
17
18 # filter
19 data = data.query("time_s_>_120")
20
21
22 # set outlier values to NaN
23 data.loc[data["id"].str.match(
24     ".*((029)|(044)|(046)|(019)|(024)|(024)|(026)|(009)|(021))"
25     ), "vicat_set_min"] = np.NaN
26
```

---

1 https://pypi.org/project/gplearn/

2 We assume, that there is *one* model to be determined for each of the responses or outputs. This is based on the idea that it is the composition of the formulation *causing*, e. g., the observed pull-off force that causes the determined thermal characteristics. Sure, there might be a correlation between the observed characteristics, but the root cause is to be found in the composition of the formulation.

```python
27  data.loc[data["id"].str.match(
28      ".*((040)|(041)|(050)|(052)|(031)|(008)|(012)|(011)|\
29  (029)|(015)|(020)|(044)|(060)|(004))"
30      ), "time_s"] = np.NaN
31
32  data.loc[data["id"].str.match(
33      ".*((023)|(028)|(033)|(024)|(020)|(030))"
34      ), "pull_off_force_N"] = np.NaN
35
36  data.loc[data["id"].str.match(
37      ".*((006)|(016)|(031)|(032)|(060)|(025)|(048)|(012)|\
38  (018)|(023)|(028)|(010))"
39      ), "viscosity_mpas"] = np.NaN
40
41
42  # copy
43  data_all = data.copy()
44
45  # %% uotputs overview
46
47
48  last_n = 4
49
50  # make boxplot
51  sns.boxplot(
52      data=data.iloc[:,-last_n:]/data.iloc[:,-last_n:].max(),
53      orient="horizontal"
54      )
55
56
57  # %% build  models
58
59  from sklearn.model_selection import train_test_split
60  from sklearn.linear_model import Ridge
61
62  from gplearn.genetic import SymbolicTransformer
63
64  list_of_summary_dicts = []
65  list_of_predictions = []
66
67  for factor_columns in [
68              ['A', 'B', 'C']
69              ]:
70      for target_column in [
71              # 'viscosity_mpas',
72              'pull_off_force_N',
73              'time_s',
74              # 'normalized_heat_flow_w_g',
75              # 'vicat_set_min'
76              ]:
77
78          # info
79          print(f"\n__===__Evaluating_for_{target_column}__===__")
80
81          # use use data from backup
82          data = data_all[~data_all[target_column].isna()]
83          # use subset only
84          data = data.iloc[:25,:] # MAX 30
85
86          # info
```

```
87      print(f"==>_Considering_{len(data)}_expriments")
88
89      # sort "out"
90      X = data[factor_columns]
91
92      # "response selection"
93      y = data[target_column]
94
95      # "manual split" in training and test data
96      test_regex = ".*[5]$"
97      test_regex_col = "id"
98
99      # select
100     X_test = data[data[test_regex_col].str.match(
101         test_regex)][factor_columns]
102     X_train = data[~data[test_regex_col].str.match(
103         test_regex)][factor_columns]
104     y_test = data[data[test_regex_col].str.match(
105         test_regex)][target_column]
106     y_train = data[~data[test_regex_col].str.match(
107         test_regex)][target_column]
108
109     # as taken from the gplearn tutorial for
110     # SymbolicTransformer:
111     # key idea: apply symbolic transformer and the apply
112     # the "Ridge" regression from sklearn
113     est = Ridge()
114     est.fit(X_train, y_train)
115     score_no_trans = est.score(X_train, y_train)
116     print("Score_without_transform:", score_no_trans)
117
118     # define function set
119     function_set = ['add', 'sub', 'mul', 'div', 'sqrt',
120                     'log', 'abs', 'neg']
121
122     # initialize the transformer
123     gp = SymbolicTransformer(
124             generations=20,   # 20
125             population_size=2000,   # 2000
126             hall_of_fame=100,
127             n_components=10,
128             function_set=function_set,
129             parsimony_coefficient=0.0005,   #"auto",   #0.0005,
130             max_samples=0.9,
131             verbose=1,
132             random_state=0,
133             #n_jobs=3
134             )
135
136     # "train" the symbolic transformer with training data
137     gp.fit(X_train, y_train)
138
139     # apply trained transformer
140     gp_freatures = gp.transform(X_train)
141     new_data = np.hstack((X_train, gp_freatures))
142
143     est = Ridge()
144     est.fit(new_data, y_train)
145     score_trans = est.score(new_data, y_train)
146     print("Score_with_transform:", score_trans)
```

```
147
148        #
149        # apply to "test" data
150
151        # transform test set
152        X_test_transformed = np.hstack((
153            X_test,
154            gp.transform(X_test)
155            ))
156
157        pred = est.predict(X_test_transformed)
158
159        plt.scatter(y_test, pred)
160        plt.xlabel("experimentally_measured")
161        plt.ylabel("predicted")
162        plt.plot([0,1], [0,1], transform=plt.gca().transAxes)
163        plt.title(target_column)
164        # show
165        plt.show()
166
167        # deviations as barplot
168        predictions = pd.DataFrame({
169            "test" : y_test,
170            "pred" : pred,
171            "sample" :  data.loc[list(y_test.index),
172                                      test_regex_col]
173            })
174
175        predictions.plot(
176            kind="barh"
177            )
178
179        plt.yticks(
180            range(len(predictions.index)),
181            predictions["sample"]
182            )
183
184        plt.title(target_column)
185        # show
186        plt.show()
187
188        # build summary dict
189        summary_dict = {
190            "factor_columns" : factor_columns,
191            "target_column" : target_column,
192            "score_no_trans" : score_no_trans,
193            "score_trans" : score_trans,
194            "predictions" : predictions,
195            "symbolic_transformer" : gp,
196            "ridge_estimator" : est
197            }
198
199        # append to list
200        list_of_summary_dicts.append(summary_dict)
201
202        # add columns to predictions df
203        predictions["target_column"] = target_column
204        predictions["factor_columns"] = "_//_".join(factor_columns)
205
206        # append to list
```

```
207           list_of_predictions.append(predictions)
208
209
210    # list of dicts to dataFramae
211    compiled_results_df = pd.DataFrame.from_records(
212                            list_of_summary_dicts)
213
214    # list of DataFrames to overall df
215    compiled_predictions = pd.concat(list_of_predictions)
216
217    # save
218    compiled_results_df.to_pickle("compiled_results.pcl")
219
220    # save
221    compiled_predictions.to_excel("compiled_results.xlsx")
```

In order to assess the quality of the resulting model, there are basically two modes of operation, both of which rely on real experimental data (see Figure 37.2).

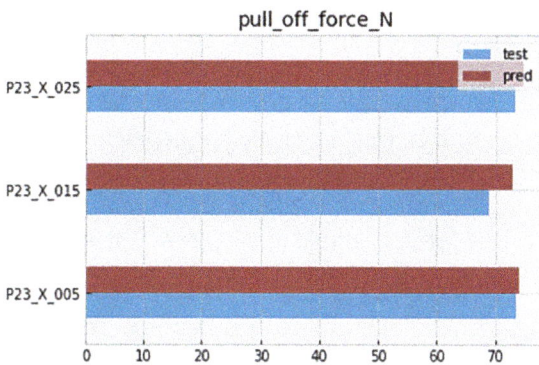

Figure 37.2: Comparison of experimental and gplearn-predicted values of pull-off force.

- Use only a part of the experimentally available data for building the gplearn model and use the observations *unknown* to the model for validation. This has the disadvantage of not using all the experimental evidence to build the best possible model. However, we get an estimate of the quality of the model based on *existing* experimental data.[3]
- Use all experimentally available data to build the gplearn model. This will give you the best possible model based on the all the experimentally available evidence. Us-

---

3 This approach is used in the code snippet presented. This is also the recommended option for modeling historical data in case no further experiments can be performed due to unavailability of raw materials or the like.

ing this approach, there is no possibility to assess the model in terms of prediction quality.[4]

The latter approaches differ only in the time at which validation experiments are available. Either they are already known at the time of modeling (scenario 1) or they are known only after additional experiments (scenario 2). Interestingly, this is a matter of choice – at least for projects currently in progress.

Discussion

Since the symbolic regression approach used in `gplearn` is a machine learning technique, the quality of the model and the corresponding agreement with actual observed values will increase with the amount of data available, i. e., observations. Therefore, it is recommended to keep track of certain quality metrics over the course of the project. A basic one could be the deviation of the actual measured values from the predicted value of a model taking into account all previous observations.

After a certain number of experiments, this value is expected to equal a certain value, ideally zero. The quality of your model does not change any more. It does not get worse, nor does it get better with additional experiments. In other words: You will not gain more insight within experimental error by conducting more experiments. You've learned what there is to learn in the system, given the available equipment. If you reach this point "too early", it's probably time to reconsider the ability of your analysis tools to resolve changes in the observed characteristics induced by varying the inputs.

Further reading

Detailed documentation of the `gplearn` package, including basic and advanced use cases, is available online.[5]

---

4 This approach is recommended for "active" projects, as an evaluation of model quality can be obtained by comparing predicted experimental results with actual results for the next experiments to be performed.

5 https://gplearn.readthedocs.io/en/stable/

38 Playing with the model or "what if"

Problem

You have an existing model (see Concept 37 of generating a model using `gplearn`) relating your inputs (such as relative component contents, processing parameters, and storage conditions) and outputs (responses such as characteristics considered relevant to your research question), and you want to obtain predictions of characteristics *assuming* a given combination of inputs, supported by existing experimental evidence.

Context

Sticking with the model of a bridge "connecting" the inputs and outputs of *conducted* experiments introduced in Concept 37, here we use it in a different manner. Assuming that the model is valid,[1] we can get predictions of the outputs to be expected given a certain set of inputs, as indicated in Figure 38.1.

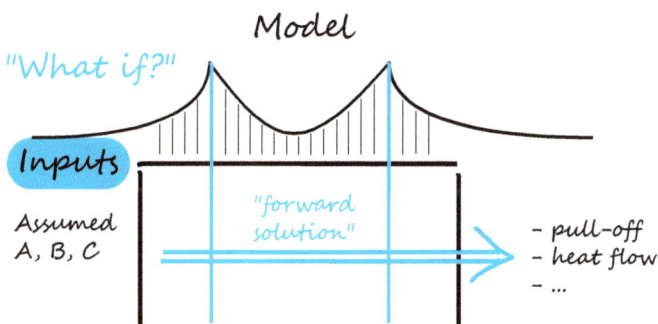

Figure 38.1: Using the model to obtain predicted values of outputs. The approach can be used to identify the most promising candidates to prepare and characterize in the lab.

But why is this relevant? The advantage of using the model to "obtain" results compared to the traditional approach of going to the lab is *not* having to go to the lab and saving considerable amounts of time spent and raw materials used. Rather than performing these experiments in reality, let the model do the work based on already existing experimental data. In real life, this approach can be used in a number of ways. For example, if you only have a limited amount of raw materials to run three more experiments, but you have an initially planned set of ten experiments remaining, you need to prioritize. Why not use the predicted characteristics of these ten formulations to select

---

1 What else can we do, right?

https://doi.org/10.1515/9783111334608-046

those for the actual experiments that result in the most desired characteristics? In this sense, model experiments and real experiments can *and* should be used hand in hand.

Solution

Use the previously generated model (see Concept 37) saved to a *pickle* file and define a function to obtain a prediction of a response value given a defined relative composition of the formulation. As shown in code section 38.1, the function get_prediction_for is therefore defined. Calculated predictions can be used either to compare a predicted value with its corresponding truly available experimental counterpart, if any, or to generate a "map" of expected, i. e., predicted values within the given input range.[2]

Listing 38.1: Predicting experimental responses, i. e., outputs given a previously generated model based on your experimental data.

```
 1  import pathlib
 2  import pandas as pd
 3
 4  # %% get raw data
 5  #
 6
 7  # define base path to data
 8  data_base_path = pathlib.Path().cwd().parent.parent \
 9      / "_Data_and_code"
10
11  # get model
12  models = pd.read_pickle("compiled_results.pcl")
13
14  # get experiments and models
15  experiments = pd.read_excel("compiled_results.xlsx")
16
17  # get compositions
18  compositions = pd.read_excel(
19      data_base_path / "_data" / "_cleaned" \
20      / "data_and_parameters.xlsx",
21  )
22
23  # merge
24  experiments = pd.merge(
25      experiments,
26      compositions,
27      left_on="sample",
28      right_on="id",
```

---

2 This is particularly tricky when working with higher dimensional formulations that contain multiple components. At this point, prior experience and/or expert knowledge is helpful in making sense of the suggested predictions. Relevant questions to ask here might be: Do the model predictions match my expectations? Do the predictions make sense taking into account my previous observations? What can I learn from the prediction/visualization? The complementary approach of using expert knowledge and modeling is highly recommended.

```
29        how="left",
30        # how="outer"
31        )
32
33
34   # %% build grid
35
36   import numpy as np
37
38   # rough
39   # fracts = np.arange(-0.001, 1.001, .1)
40   # fine
41   fracts = np.arange(-0.001, 1.001, .05)
42   # # very fine
43   # fracts = np.arange(-0.001, 1.001, .025)
44   # # very very fine
45   # fracts = np.arange(-0.001, 1.001, .01)
46
47   # init list
48   concs = []
49
50   for f in fracts:
51       for i in range(len(fracts)):
52           concs.append(f)
53
54   # init list
55   concsII = []
56
57   for i in range(len(fracts)):
58       for f in fracts:
59           concsII.append(f)
60
61   # init DataFrame
62   grid = pd.DataFrame()
63
64   # fill DataFrame
65   grid["A"] = concs
66   grid["B"] = concsII
67   grid["C"] = 1 - grid["A"]- grid["B"]
68
69   # filter
70   grid = grid.query("C_>=_0")
71
72   # clean
73   del fracts, concs, concsII, i, f
74
75
76   # %% model function
77
78   from sklearn.utils import check_array
79
80   def get_prediction_for(pcl, model_nr, experimental_settings,
81                          show_info=True):
82       """
83       get a prediction for a gplearn model given defined
84       experimental settings, i.e., formulation composition
85       """
86
87       # info
88       if show_info:
```

```
89          print("predicting", pcl.at[model_nr, "target_column"])
90          print("with_input_for", pcl.at[model_nr, "factor_columns"])
91
92      # convert to Series
93      experimental_settings = pd.Series({
94              "A" : experimental_settings[0],
95              "B" : experimental_settings[1],
96              "C" : experimental_settings[2]
97              })
98
99      # get transformer
100     transformer = pcl.at[model_nr, "symbolic_transformer"]
101
102     # "harvest" input
103     x_list = []
104     for entry in pcl.at[model_nr, "factor_columns"]:
105         # get input
106         x_list.append(experimental_settings[entry])
107
108     # type conversion
109     X = check_array(np.array([x_list], dtype=float))
110
111     # transform
112     X_transformed = np.hstack((
113                         X,
114                         transformer.transform(X)
115                         ))
116
117     # get estimator
118     estimator = pcl.at[model_nr, "ridge_estimator"]
119
120     # estimate
121     prediction = estimator.predict(X_transformed)[0]
122
123     # return
124     return float(prediction)
125
126
127 # %% options
128
129
130 # # target = "viscosity_mpas"
131 # # target = "characteristic_vicat_time_min"
132 target = "pull_off_force_N"
133 target = "time_s"
134
135 # get model number based on target characteristic name
136 model_nr = int(models[models["target_column"]
137                 == target].index.values[0])
138
139
140 # %% prediction for one composition
141
142 # define formulation
143 a, b, c = 0.4, 0.15, 0.45
144
145 # info
146 print(f"Getting_prediction_for_a={a},_b={b}_and_c={c}.")
147
148 # get predicted value
```

```python
149  prediction = get_prediction_for(
150      pcl=models,
151      model_nr=model_nr,
152      experimental_settings=[a, b, c],
153      show_info=False
154      )
155
156  # show results
157  print(f"==>_Calculated_{target}:_{prediction:.2f}.")
158
159
160  # %%
161
162  predictions = []
163  # get predictions for each combination of components A, B
164  # and C in the "formulation grid"
165  for a, b, c in zip(grid.A, grid.B, grid.C):
166      predictions.append(
167              get_prediction_for(
168                  pcl=models,
169                  model_nr=model_nr,
170                  experimental_settings=[a, b, c],
171                  show_info=False
172              )
173          )
174
175  del a, b, c
176
177  # append
178  grid["target"] = predictions
179
180  # %%
181
182  import matplotlib.pyplot as plt
183  import mpltern
184
185  # filter data
186  grid = grid.query("target_>=_0")
187
188  grid = grid.query("A_>=_0.1")
189  grid = grid.query("B_<=_0.5")
190  grid = grid.query("C_>=_0.15")
191
192  # select experiments
193  experiments_selected = experiments[
194          experiments["target_column"]==target]
195
196  # define colormap
197  cmap = "inferno"
198
199  # init figure
200  fig = plt.figure(figsize=(10, 5))
201  ax = fig.add_subplot(projection='ternary')
202
203  # deine "levels"
204  levels = np.linspace(
205      grid["target"].min(),
206      grid["target"].max(),
207      6   # num
208      )
```

```
209
210   # text
211   for i, row in experiments_selected.iterrows():
212       ax.text(
213           row["A"],
214           row["B"],
215           row["C"],
216           f'{row["pred"]:.0f}_s__',
217           ha='right', va='center'
218           )
219
220
221   cs = ax.tricontourf(
222       grid.A,
223       grid.B,
224       grid.C,
225       grid.target, # color code by
226       levels=levels,
227       alpha=0.8,
228       vmin=levels.min(),
229       vmax=levels.max(),
230       cmap=cmap
231       )
232
233   # scatter plot
234   ax.scatter(
235       experiments_selected["A"],
236       experiments_selected["B"],
237       experiments_selected["C"],
238       c=experiments_selected["pred"],
239       vmin=levels.min(),
240       vmax=levels.max(),
241       edgecolors="black",
242       cmap=cmap
243       )
244
245   # set labels
246   ax.set_tlabel("A")   # "top"
247   ax.set_llabel("B")   # "left"
248   ax.set_rlabel("C")   # "right"
249
250   # # set limits
251   # ax.set_tlim(0.25, 1)   # "top"
252   # ax.set_rlim(0.2, 1)   # "right"
253
254   # grids
255   ax.grid(axis='t', color="gray")
256   ax.grid(axis='l', which='minor', linestyle='--', color="gray")
257   ax.grid(axis='r', which='both', linestyle=':', color="gray")
258
259   # add color bar
260   cax = ax.inset_axes([1.05, 0.1, 0.05, 0.8],
261                         transform=ax.transAxes)
262   colorbar = fig.colorbar(cs, cax=cax)
263   colorbar.set_label(target, rotation=270, va="baseline")
264
265   # save
266   plt.savefig(
267       "ternary.png",
268       bbox_inches="tight",
```

```
269    dpi=300
270    )
```

The predicted values of the response variable are shown in the ternary plot of Figure 38.2.

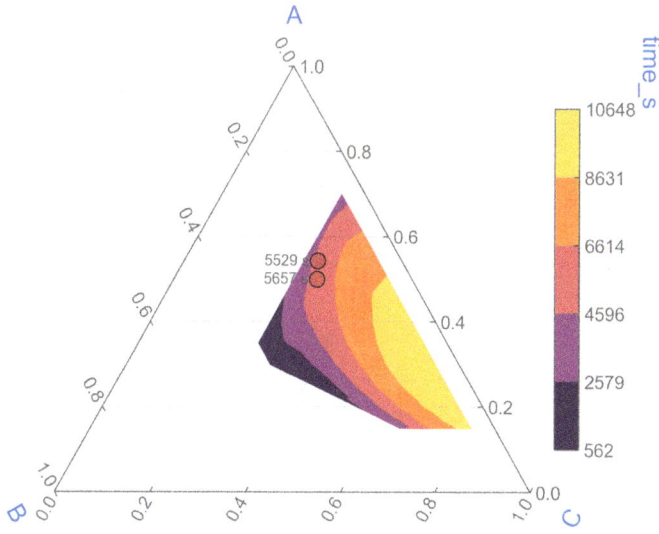

Figure 38.2: Experimentally observed values of the characteristic *peak position* shown next to the predicted values as obtained from the previously derived gplearn model.

Try some of the indicated options (coarse, fine, very fine, very very fine) for grid spacing in the ternary plot by removing the corresponding comment in code section 38.1. Additionally, try using the get_prediction_for for another of the available model_nr options.

Discussion

There is little to say about the complementary approach of using model predictions and physically conducted experiments.
- Predictions of characteristics can be used to select the "most promising" candidates based on the experimental evidence available to date.
- The deviation between the predicted value of a feature and its actual experimental value can be understood as a measure of model quality.

- Additionally collected experimental evidence should be included to the experiment to obtain a better description of experimental reality.[3]
- Consider the prediction of characteristics related to a composition / process combination and validating them through experimentation as an iterative process. Ideally, the deviation between predictions and experimental values should decrease with increasing number of experiments.[4]

The concept of iterative prediction, running the proposed experiments, and recalculating the model is outlined in Figure 38.3.

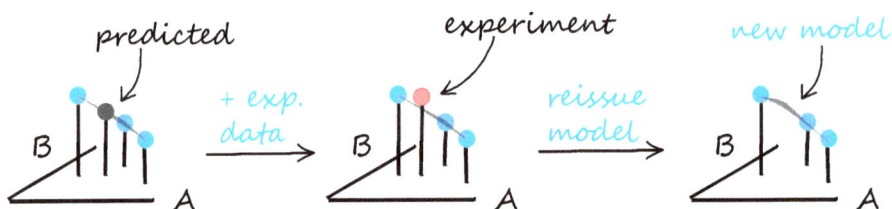

Figure 38.3: Conceptual sketch of iterative model building, predicting experimental results, conducting these experiments to prove/falsify the model predictions, and incorporating the new experimental values into the model.

Further reading

The same principle is also used in Concept 39, where the visualization is achieved using a *jupyter notebook*.

---

3 Keep in mind the famous statement of George Box: "*All models are wrong, some are useful.*".

4 Assuming a reasonably chosen formulation space, i. e., range of inputs and responses "of good nature", i. e., free of step-like or other types of unconventional behavior. To account for these latter behaviors, the input space could be adjusted accordingly, or the multiple models could be generated.

39 Playing with the model or – `jupyter` notebook

Problem

You want to obtain predictions from an experimentally backed model using a Graphical User Interface (GUI) to improve your understanding of the system being analyzed.

Context

Creating a data-driven model is an essential part of any modern R&D project that aims to shed light on the interaction of mixed components. As indicated in Concept 38, "playing" is an integral part of understanding the implications for real-world applications. Therefore, a readily accessible way to generate predictions based on the model other than explicitly typing in the relative contents of a formulation for which a particular characteristic is to be predicted.

Solution

Use the `notebook`-package[1] as a web-based notebook environment for interactive computing with Python. It can be installed with `pip` via `pip install notebook` and started from the console with the command `jupyter notebook`.

Within these interactive notebooks, basic GUIs (see Figure 39.1) can be easily built to define, e. g., the relative compositions of a formulation using sliders.[2]

In code section 39.1, the creation of a basic input GUI is shown next to the calculation and basic visualization of a calculated model result. For simplicity, the raw material related cost of the formulation is calculated based on the relative contents and a defined price of the components.[3]

Listing 39.1: Building a basic GUI in `jupyter notebook`.

```
1  import ipywidgets as widgets
2  from ipywidgets import interactive
3  import matplotlib.pyplot as plt
4
5  a = widgets.FloatSlider(value=.40, min=0.40, max=0.65, step=0.01,
```

1 https://pypi.org/project/notebook/

2 A list of available widgets is accessible here: https://ipywidgets.readthedocs.io/en/stable/index.html. There are sliders for float and integer inputs, progress bars, text input and output fields, checkboxes and buttons, among others.

3 Instead of the basic "price" model, any other response model could be used. Price was chosen for convenience in this example.

https://doi.org/10.1515/9783111334608-047

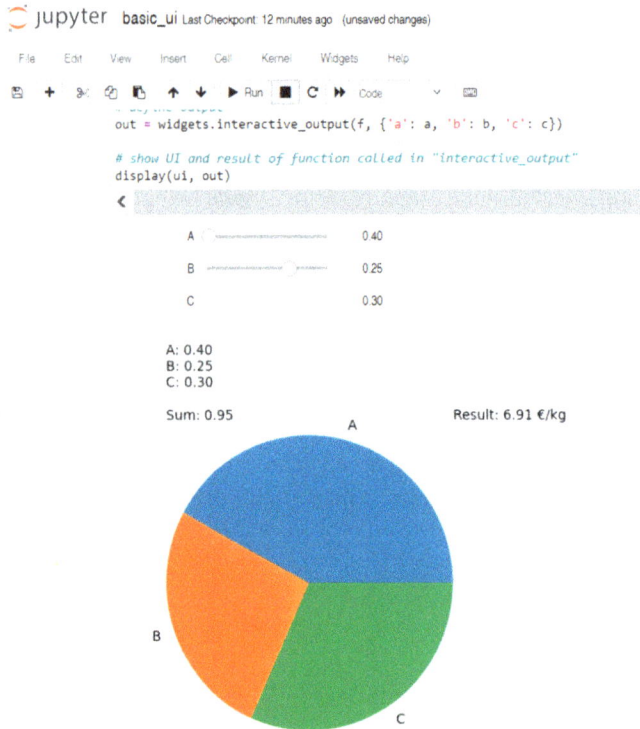

Figure 39.1: Screenshot of the resulting GUI obtained by `jupyter notebook`.

```
 6                     description='A', readout=True, readout_format='.2f')
 7  b = widgets.FloatSlider(value=.25, min=0.04, max=0.35, step=0.01,
 8                     description='B', readout=True, readout_format='.2f')
 9  c = widgets.FloatSlider(value=.30, min=0.00, max=0.40, step=0.01,
10                     description='C', readout=True, readout_format='.2f',
11                     disabled=True)
12
13  # calculate value of C from A and B
14  def update_c(*args):
15      c.value = 1 - a.value - b.value
16
17  # update C value based on A and B
18  a.observe(update_c, 'value')
19  b.observe(update_c, 'value')
20
21  # stack components in vertical box
22  ui = widgets.VBox([a, b, c])
23
24  # define function to be called on interaction with UI
25  def f(a, b, c):
26
27      # calculate model here
28      model_value = 4*a + 3*b + c/a
29
30      # get checksum
```

```
31      sum = a + b + c
32
33      # plot
34      plt.pie(
35          [a, b, c],
36          labels=["A", "B", "C"]
37      )
38
39      # composition info
40      plt.text(-1, 1, f"A:_{a:.2f}\nB:_{b:.2f}\n\
41  C:_{c:.2f}\n\nSum:_{sum}\n")
42
43      # result info
44      plt.text(1, 1, f"Result:_{model_value:.2f}\n")
45
46  # define output
47  out = widgets.interactive_output(f, {'a': a, 'b': b, 'c': c})
48
49  # show UI and result of function called in "interactive_output"
50  display(ui, out)
```

Three sliders a, b and c have been created to define the composition of the formulation. As a boundary condition, the content of component C is calculated to complement components A and B to a relative content of 1. Every time one of the sliders is moved, the material cost associated with that composition is calculated as a simplified model function.

Replace the trivial model in the code above with the model used in Concept 38.

Discussion

As mentioned above, jupyter notebooks are a widely used and appreciated way of using Python interactively via a GUI. Code is run within cells, plots are displayed right below the defining code and the entire script can be enhanced via Markdown to add further details, though processes and information that make the content easier to understand at a later point in time. Also, jupyter notebook is common for learning purposes, starting with the journey in the Python world.[4]

To run the jupyter notebook file directly, the command jupyter notebook path_to_file can be run from the command line.

---

4 https://www.learnpythonwithjupyter.com/

Further reading

If you prefer to use conventional Python files instead of a `jupyter notebook`, the latter can be converted into a Python file with the following command:

```
jupyter nbconvert --to script *.ipynb
```

40 Playing with the model or – `voila`

Problem

You want to obtain predictions from an experimentally validated model using a standalone GUI to increase your understanding of your system.

Context

As stated in the previous Concept 39, creating and "playing" with a model is an essential part in R&D projects. Having a *standalone* GUI at hand just adds a further layer of convenience on top of the previous state.

Solution

Use the `voila` package[1] to turn an interactive `jupyter` notebook as described in Concept 39 into standalone applications that run in the web browser.

To run the standalone application in the browser, execute

```
1  voila path\_to\_file
```

in the command line.

Alternatively, the application can also be launched using a batch-script as outlined in code section 40.1.[2]

Listing 40.1: Run a jupyter notebook as a standalone application using `voila`.

```
1   @echo OFF
2   rem https://gist.github.com/maximlt/531419545b039fa33f8845e5bc92edd6
3
4   set CONDAPATH=C:\Users\MatthiasHofmann\anaconda3
5   set ENVNAME=base
6
7   if %ENVNAME%==base (set ENVPATH=%CONDAPATH%)^
8    else (set ENVPATH=%CONDAPATH%\envs\%ENVNAME%)
9
10  call %CONDAPATH%\Scripts\activate.bat %ENVPATH%
11
12  REM Changing directory to path of jupyter nptebook
13  cd "C:\Users\MatthiasHofmann\Documents\GitHub\^
14  data_playbook_II\_Manuscript\074"
```

---

1 https://pypi.org/project/voila/

2 see https://github.com/itsergiu/voila/blob/main/start_voila_in_conda_env.bat

https://doi.org/10.1515/9783111334608-048

```
15
16  REM Call notebook via voila
17  voila basic_ui.ipynb
18
19  REM Keeping console open
20  pause
```

The GUI displayed in the browser is shown in Figure 40.1. The sliders in the top section serve to set the relative contents of the formulation components A and B. As in the previous concept, component C is calculated to complement the formulation. As the sliders are moved, a simplistic model for the raw material cost is also created.[3]

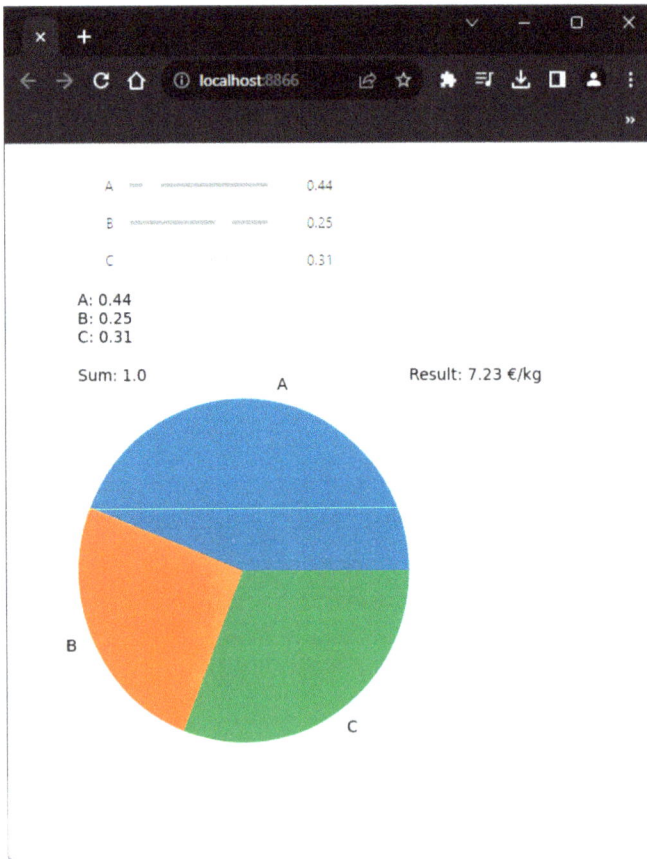

Figure 40.1: Screenshot of the standalone `voila` application in the browser.

---

3 Even though not immediately obvious to the thought process of natural scientists, the cost associated with the composition of a formulation can be considered a model. It might feel unusual as there's no un-

Corresponding to the selected composition it is calculated and shown as textual information.

Discussion

Using `voila` on top of juypter notebooks adds a layer of convenience for *using* a GUI without additional clutter. It allows for having a clean interface and focus on the actual (scientific) content of your work. It can be considered as one of the "later stages" of working with your experimental data. Only when you know exactly what should be displayed, it makes sense to build this type of interface. Using `voila`, however, you are somewhat further away from your data. In this sense, its use can be considered a trade-off and its benefits should be evaluated on a case-by-case basis.

Further reading

A gallery of usage examples for `voila` is available online.[4]

---

certainty involved in calculating a price when unit costs and relative contents are known. It this sense, the cost model can be considered superior to the natural science property models in terms of uncertainty. Conceptually, however there's a similarity in mapping the relative contents to *a* characteristic – the scientific models are typically just somewhat more intricate.

4 https://voila-gallery.org/

41 Playing with the model or – `streamlit`

Problem

You want to get predictions from a model (supported by experimental determined findings) using a standalone GUI to increase your understanding of your system.

Context

As stated in the previous concept 39, creating and "playing" with a model is an essential part in R&D projects. Having a *standalone* GUI at hand just adds another layer of convenience on top of the previous state.

Solution

Use the `streamlit`-package[1] to build and share applications based on your (scientific) data.

To install via `pip`, use the command `pip install streamlit`. The success of the installation can be verified with the command `streamlit hello`, which opens up the introductory application shown in Figure 41.1 in the browser.

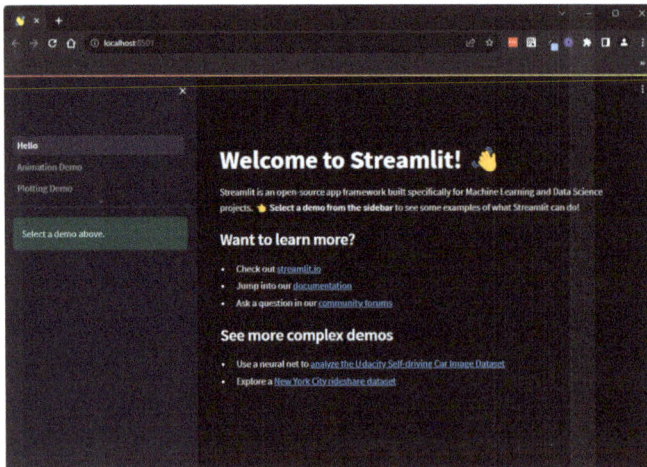

Figure 41.1: The application displayed in the browser upon calling `streamlit hello`.

---

1 https://pypi.org/project/streamlit/

https://doi.org/10.1515/9783111334608-049

As a next step, we build the script including the widgets provided by the `streamlit` package. When a slider is moved, button clicked or the user interacts in any other way with the interface, the entire script is re-run.[2] To call the script and open the application in the browser, we can run

```
1 streamlit run path\_to\_file
```

in the command line. As in the case of the voila-app (see Concept 40), a `streamlit` GUI can be started using a batch-script as shown in code section 41.1. The first step is to activate the desired Python environment, change to the directory where the `streamlit` application is located and call it with the command `streamlit run` with the name of the file as argument.

Listing 41.1: Running a `streamlit` application using a *batch*-script.

```
1  @echo OFF
2  rem https://gist.github.com/maximlt/531419545b039fa33f8845e5bc92edd6
3
4  set CONDAPATH=C:\Users\MatthiasHofmann\anaconda3
5  set ENVNAME=base
6
7  if %ENVNAME%==base (set ENVPATH=%CONDAPATH%)^
8   else (set ENVPATH=%CONDAPATH%\envs\%ENVNAME%)
9
10 call %CONDAPATH%\Scripts\activate.bat %ENVPATH%
11
12 REM Changing directory to path of the python (.py) file
13 cd "C:\Users\MatthiasHofmann\Documents\GitHub\^
14 data_playbook_II\_Manuscript\076"
15
16 REM Call the python script via streamlit
17 streamlit run code.py
18
19 REM Keeping console open
20 pause
```

The resulting app includes some coupled sliders to define the relative composition of our ternary formulation consisting of components A, B and C, similar to the one shown in Concept 40. The corresponding code is given in code section 41.2.

Listing 41.2: Source code for the `streamlit` GUI.

```
1 import streamlit as st
2 import pandas as pd
3 import plotly.express as px
4
```

---

2 For time-consuming operations it is possible to use the caching mechanism provided by `streamlit`.

```
5   # add title
6   st.title("Cost_Modelling_for_Example_Formulation")
7
8   # calculate value of C from A and B
9   def update_c(*args):
10      # calculate relative content of compoennt C
11      calc_value = 1 - st.session_state.a - st.session_state.b
12
13      # set new value of slider "c"
14      st.session_state["c"] = calc_value
15
16  # add horizontal role divider
17  st.divider()
18
19  # add Info text
20  """This part of the formulation serves to define the
21  *relative* contents of components A, B and C in the
22  formulation. The contente of C is calculated to
23  complement the formuation to an overall content of 1.
24  """
25
26  # make sliders
27  a = st.slider("A", value=0.45, min_value=0.40, max_value=0.65,
28                key="a", on_change=update_c)
29  b = st.slider("B", value=0.25, min_value=0.04, max_value=0.35,
30                key="b", on_change=update_c)
31  c = st.slider("C", value=0.30, min_value=0.00, max_value=0.40,
32                key="c", disabled=True)
33
34  # intialize session states
35  if "a" not in st.session_state:
36      st.session_state["a"] = a
37
38  if "b" not in st.session_state:
39      st.session_state["b"] = b
40
41  if "c" not in st.session_state:
42      st.session_state["c"] = c
43
44  # write text
45  st.write(f"Sum_of_components_A,_B_and_C:_{a_+_b_+_c:.2f}")
46
47  if a + b + c == 1:
48      # success box
49      st.success("Components_add_up_to_1!")
50  else:
51      # error box
52      st.error("Components_do_not_add_up_to_1!")
53
54  # add horizontal role divider
55  st.divider()
56
57  # caption
58  st.caption("Calcuation")
59
60  # add Info text
61  """
62  In this section, the theoretical cost assuming a certain unit
63  cost scenario (see sidebar for selection) and taking into account
64  the selected formulation is calculated.
```

```
65  """
66
67  # add options to sidebar
68  cost_option = st.sidebar.selectbox(
69      "Select a cost scenario",
70      ("optimistic", "pessimistic"),
71      index=1  # select "pessimistic" scenario as default
72      )
73
74  # define structure for unit costs (for simplicity: dict of dicts)
75  unit_cost = {
76      "optimistic" : {"a" : 2.0, "b": 3.7, "c" : 5.3},
77      "pessimistic" : {"a" : 2.9, "b": 5.7, "c" : 58.3},
78      }
79
80  # calculate cost (== use a "model")
81  this_scenario_cost =\
82      a * unit_cost[cost_option]["a"] +\
83      b * unit_cost[cost_option]["b"] +\
84      c * unit_cost[cost_option]["c"]
85
86  # Info text
87  f"""
88  Assuming the **{cost_option}** scenario, the chosen formulation
89  is connected to a unit cost of {this_scenario_cost:.2f} Euro/kg.
90  """
91
92  # add horizontal role divider
93  st.divider()
94
95  # caption
96  st.caption("Visualization")
97
98  # add Info text
99  """
100 This section visualizes the chosen composition and resulting cost.
101 """
102
103 # build dataframe for visualization
104 data = pd.DataFrame.from_dict({
105     "component" : ["A", "B", "C"],
106     "relative_content" : [a, b, c],
107     "price_contribution": [
108         a * unit_cost[cost_option]["a"],
109         b * unit_cost[cost_option]["b"],
110         c * unit_cost[cost_option]["c"]
111         ]
112     })
113
114 # calculate relative price contribution
115 data["relative_price_contribution"] =\
116     data["price_contribution"] / data["price_contribution"].sum()
117
118 # Define tabs
119 tab_comp, tab_price = st.tabs(["Composition", "Pricing"])
120
121
122 with tab_comp:
123     # make plotly plot
124     fig = px.pie(
```

```
125          data,
126          values="relative_content",
127          names="component",
128          hover_data=["relative_price_contribution"],
129          hole=0.4,   # make pie chart a donut chart
130          title="Relative_Formulation_Contents",
131          color='component',
132          color_discrete_map={
133                  'A':'lightcyan',
134                  'B':'cyan',
135                  'C':'royalblue'
136              }
137          )
138
139      # This is the default. So you can also omit the theme argument.
140      st.plotly_chart(fig, theme="streamlit", use_container_width=True)
141
142  with tab_price:
143      fig = px.pie(
144          data,
145          values="relative_price_contribution",
146          names="component",
147          hover_data=["relative_content"],
148          hole=0.4,   # make pie chart a donut chart
149          title="Relative_Pricing_Contents",
150          color='component',
151          color_discrete_map={
152                  'A':'lightcyan',
153                  'B':'cyan',
154                  'C':'royalblue'
155              }
156          )
157
158      # This is the default. So you can also omit the theme argument.
159      st.plotly_chart(fig, theme="streamlit", use_container_width=True)
160
161  # show DataFrame
162  data
```

Sticking with the trivial cost model, we now have the option to choose either an *optimistic* or *pessimistic* cost scenario from the sidebar (see Figure 41.2). In addition, we choose to visualize the formulation in terms of either relative content or relative cost contributions, depending on the user's choice.

Discussion

To make the transition from conventional scripting, which marks the entry point to the Python-world for many natural scientists, to building applications, the use of `streamlit` is highly recommended. It allows you to maintain the "script-like" nature of your code. To transition, run your script with the command

```
1  streamlit run my\_script.py
```

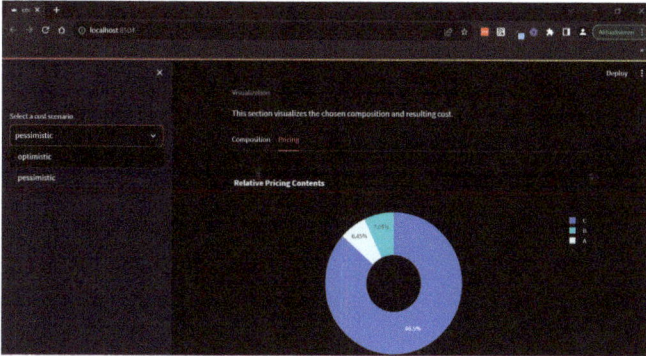

Figure 41.2: The resulting `streamlit` GUI for visualizing the component and cost composition of ternary formulations according to user input.

and wait for the browser to open. As you add interactivity feature by feature, the `streamlit` application changes and grows on reload. You want to be able to change, e. g., the relative content of a formulation's component? Just replace the so far existing fixed number variable with a slider (see code section 41.2). Similarly, changing scenarios, i. e., can be achieved by selecting from several discrete options in drop-down menus. Depending on your use case, there may be more relevant widgets for your application. Again, the extensive online documentation is a valuable source of information.[3]

Further reading

For information on deploying, managing, and sharing an application in the Community Cloud, see Concept 47.

---

3 https://docs.streamlit.io/library/api-reference/widgets

42 Dealing with too few experiments

Problem

You want to build a model based on experimental data, but there aren't enough experiments (yet).

Context

Unlike traditional *data science* topics, one of the main challenges in *science data* is the lack of – caution: buzzword! – "big data", i. e., large number of *observations*. Unlike in, e. g., purchase-related or user activity-related data, experimental data does not come incidentally. Experiments need to be thoroughly planned (with a purpose) (Concept 26 and Concept 27), carefully conducted and precisely analyzed. It's only then that you arrive at the starting point for data science-like activities. In other words: Observations, i. e., experiments are costly and typically involve a considerable amount of – paid – work time. Therefore, it is understandable that experimental high-quality experimental data is, ironically, the scarcest resource in many data science projects in the natural sciences, such as those conducted in (more or less) routine settings like industrial formulation research.

Solution

Combine the statistical *bootstrapping* method with your knowledge of the experimental processes used.

In a broader sense, bootstrapping belongs to the class of *resampling methods*. The key idea is to model the characteristics of an overall system from a pre-existing set of (experimental) data. More technically, bootstrapping assumes that the inference of a population from sample data can be modeled by *resampling with replacement* already existing sample data, as show in Figure 42.1.

Essentially, bootstrapping allows you to do more with less. In our case, we can address the issue of (always) having too few experiments at hand. Using a bootstrapping-like approach, we can obtain a better picture of the "true" relationship between experimental inputs and corresponding outcomes by building on actually conducted experiments combined with (statistical) knowledge of the experimental process. Thereby we can generate additional (you might as well call them "synthetic") experiments, which can be used in creating an improved model.

Let's dive into the question of the basic assumptions that allow us to do this. In the end, it's the realization that even chemistry and materials science are – at least in the context of the field covered and formulated here – not exact sciences. Uncertainties lurk

https://doi.org/10.1515/9783111334608-050

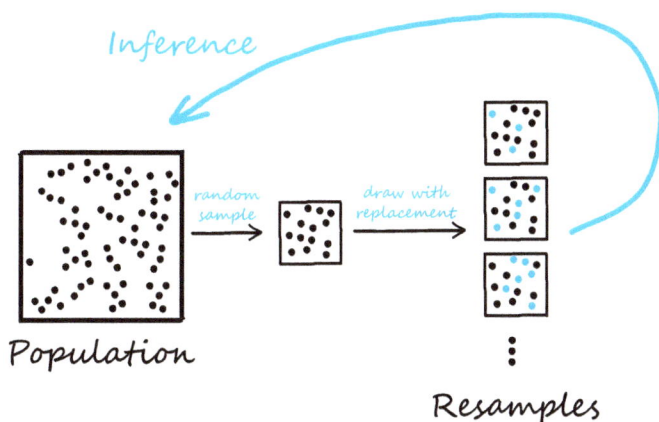

Figure 42.1: Schematic of the bootstrapping concept: Characteristics of the overall system can be inferred by resampling from an existing subset "with replacement".

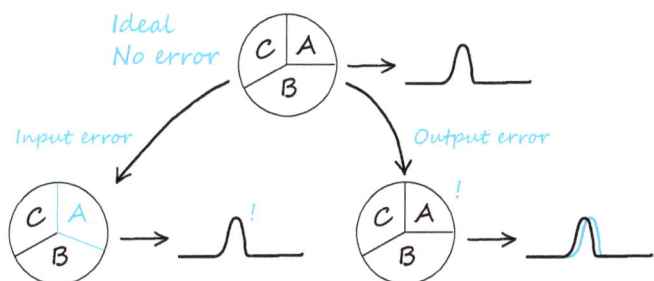

Figure 42.2: Bootstrapping as an interpretation of experimental uncertainty.

at both ends of *inputs* and *outputs* as indicated in Figure 42.2. Sources of uncertainty at both ends of the experimental chain are related to questions such as

- Were the compounds precisely weighed as planned?
- Were accurately weighed in compounds properly documented?
- Do the compounds used meet the declared specifications?
- Are the containers and equipment used free from traces of previous experiments?
- Have the materials been subjected to certain aging effects?
- Has the analytical equipment been properly calibrated and maintained?
- Has the analytical equipment been operated adequately?

These are certainly only a few of the considerations to be taken into account.[1] Now that there are all too many potential sources of error, there are essentially two ways of dealing with these circumstances. Either you want to eliminate all of them, which doesn't

---

1 Not to mention "unknown unknowns".

make much economic sense, and worse, we end up with even fewer experiments. Alternatively, we can embrace the knowledge provided by these uncertainties and leverage them to add insights to our few, but already existing experiments.

In the following we'll start with the experiments that have actually been conducted and "envision alternative endings and beginnings" for them, taking into account the uncertainties associated with either the input or the output side.

On the *input* side, uncertainties mostly arise from the precision with which a plan can be turned into reality. For simplicity, let's assume, that the specified masses are weighed in with a precision of 0.25 g. Depending on your actual processes, the uncertainties may be larger or smaller, depending on the rotational speed of the mixing devices, the storage temperatures applied and much more. Also, these limits can be increased above a certain reasonable threshold to learn more about your systems. Interestingly, this is also the part where interpersonal aspects come into play. A pessimistic person has less confidence in the work of others and may define a larger potential error range here, while a more positive person may go down to the precision specified on the scale in the example above.

As shown in code section 42.1, we select the first 15 experiments as given and want to generate additional experiments based on the *planned* weighted samples. Considering a batch size of 100 g each, we assume that the maximum individual error of each weighing step is 0.75 g. Therefore, we assume that the mass of each actual component can be lower or higher up to this value, independent of the contents of the other components. For simplicity, we will just draw a random value that increases or decreases the planned mass. In the given example, we end up with a total of 60 formulations based on the 15 experiments actually performed. The 45 generated experiments can be considered as "alternative experiments" that you might even have conducted without realizing it.

Listing 42.1: Incorporating bootstrapping in natural scientific exponents.

```
1   import pathlib
2   import pandas as pd
3   import numpy as np
4   import matplotlib.pyplot as plt
5
6   # ensure consitent random state
7   np.random.seed(seed=202309)
8
9   # define data path
10  path_data = pathlib.Path().cwd().parent.parent \
11      / "_Data_and_code" / "_data"
12
13  # get formulation composition
14  compositions_meas = pd.read_excel(
15      path_data / "_raw" / "composition" / "composition.xlsx",
16      usecols="B:E",
17      nrows=15,  # restrict to first 15 experiments
18  ).set_index("id")
19
```

```python
20  # assume formulation batches of 100g in each experiments
21  compositions_meas *= 100
22
23  # check total mass
24  compositions_meas["total_mass_g"] = compositions_meas.sum(axis=1)
25
26  # add information on origin
27  compositions_meas["source"] = "experiment"
28
29  # # "push back index"
30  # compositions = compositions.reset_index()
31
32  # define a function to pull a random(!) error
33  def get_random_uncertainty(limit : float = 1):
34      """
35      returns a random floating-point number to be used as error
36      """
37
38      # get uncertainty
39      uncertainty = np.random.uniform(-limit, limit)
40
41      # return
42      return uncertainty
43
44  # init list of dicts
45  list_dicts = []
46
47  # set(!) weigh-in maximum error
48  error_limit = 0.75  # [g]
49
50  # get possibly analyzed alternatives for each target(!) formulations
51  for i, row in compositions_meas.iterrows():
52      # number of "created" experiments per actual experiments
53      for _ in range(3):
54          # determine "alternative" possible compositions of experiments
55          _a = row["A"] + get_random_uncertainty(limit=error_limit)
56          _b = row["B"] + get_random_uncertainty(limit=error_limit)
57          _c = row["C"] + get_random_uncertainty(limit=error_limit)
58
59          # merge "alternative" composition to a row
60          _row = {"id" : i, "A" : _a, "B" : _b, "C" : _c,
61                  "source" : "generated"}
62          # append to list
63          list_dicts.append(_row)
64
65  # convert list of dicts to DataFrame
66  compositions_created = pd.DataFrame(list_dicts).set_index("id")
67
68  # calculate total mass
69  compositions_created["total_mass_g"] = compositions_created.\
70      iloc[:,:3].sum(axis=1)
71
72  # combine the dataframe
73  compositions = pd.concat([compositions_meas, compositions_created])
74
75  # calculate new relative contents
76  for comp in ["A", "B", "C"]:
77      # scale
78      compositions[comp] /= compositions["total_mass_g"]
79
```

```
80   # plot
81   for sample, sample_data in compositions.groupby(by="id"):
82       # set source as index column
83       sample_data = sample_data.set_index("source")
84       # restrict to columns A, B and C
85       sample_data = sample_data.iloc[:,:3]
86
87       # show experiments
88       p = sample_data.plot(kind="line", marker="o")
89
90       # add guide to the eye lines
91       for i in sample_data.loc["experiment",:]:
92           # add horizontal line
93           plt.axhline(i, linewidth=0.75, alpha=0.75, color="k")
94
95       # add label
96       plt.ylabel("Relative_content_/_[1]")
97
98       # add title
99       plt.title(f"Experiments_based_on_{sample}")
100
101      # rotate x-ticks
102      plt.xticks(rotation=70)
103
104      # set figure size
105      plt.gcf().set_size_inches(4,8)
106
107      # show plot
108      plt.show()
109
110      break
```

Adjust the code to create more, e. g., 30 or 1000 generated experiments based on the initial 15 actually performed experiments. How does the plot corresponding to Figure 42.3 change? Does this match your intuition as a natural scientist?

In the code above, only the input side has been considered. Uncertainties on this side arise from either the (specified) (im)precision of the devices involved, or the precision of the execution of a plan – in this sense a human factor. Therefore, the maximum uncertainty range has to be *assumed* to a certain extent rather than it can be precisely measured.

On the other hand, the *outputs*, i. e., extracted characteristics (such as peak positions, viscosities, etc.), the maximum error relevant for the calculation of uncertainties can be obtained by repeating a *measurement* for one and the same sample multiple times.[2] Based on these replicates, a maximum error for the evaluation of uncertainties can be determined.

---

2 Certainly, this is only possible in the case of non-destructive testing and in situations where the sample does not change (significantly) between subsequent measurements, i. e., unreactive mixtures. If repeated measurements on one and the same sample are not possible in your case, you'll have to do the next best

Finally, the randomly modified inputs combined with the corresponding randomly modified outputs will give you access to numerous additional "alternative endings" of your experiments that you just did not observe for statistical reasons. But still, based on some existing experiments and some process knowledge, the problem of having too few observations at hand can be greatly alleviated.

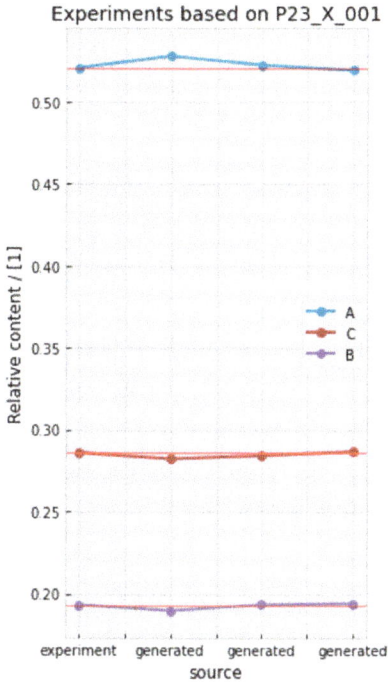

Figure 42.3: Alternative "generated" ternary formulations taking into account typical process errors based on the originally "assumed" experiment.

Discussion

The question may arise as to why (or why not) it makes sense to draw a random uniformly distributed uncertainty from a certain range in order to modify the intended input values and observed outputs. A striking reason to stick with this procedure is the recognition of the fact, that we are not certain that our *anker point*, the actually conducted experiment itself has been prepared strictly according to plan, and that the observed experimental result (or derived characteristic) is at the center of a now-uniform

thing to estimate the maximum error of your method: Prepare multiple batches of the same formulation and perform the measurements to create replicates.

distribution.[3] At least this seems to be the case in most real-world scenarios, and especially when trying to make sense of historical data in an *a posteriori* fashion.

What should be considered in this approach? First and foremost, it is critical that the "true" experimental *anker points* are sufficiently far apart from each other. Only then, the experimental methods will be able to resolve the effects induced by modifying either some aspects of the studied composition and/or the processing parameters applied. The aim of the concept outlined above is to transform an initially "hard", measured experimental value into a "cloud" of potential alternative outcomes that might have occurred taking into account both errors on the input and processing side as well as characterization errors. An outline of the process is shown in Figure 42.4.

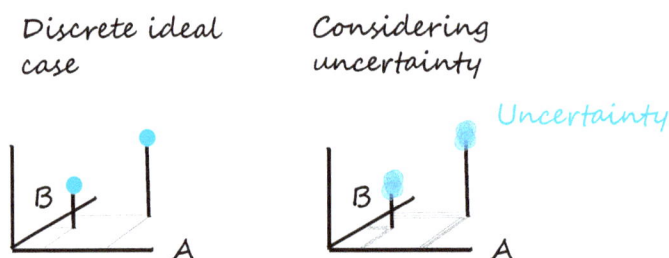

Figure 42.4: Relation between planned input and measured output. Taking into account the bootstrapping approach, the individual points are "softened" into a "cloud" of possible alternative outcomes.

Further reading

For an introduction to bootstrapping, see *Bootstrap Methods- With Applications in R*[4] and *Bootstrapping- An Integrated Approach with Python and Stata.*[5]

---

3 At least on the output side one may argue, that the observed value is subject to a non-uniform distribution, given that the inputs are precisely known and truthfully documented.

4 https://link.springer.com/book/10.1007/978-3-030-73480-0

5 https://www.degruyter.com/document/doi/10.1515/9783110693348/html?lang=en

43 Solving the reverse problem applying multiobjective optimization

Problem

You want to obtain recommendations on which and how to mix certain components to obtain derived characteristics supported by experimental data.

Context

So far, we have used our more (experimental characteristics) or less (unit cost) complex models to obtain a predicted value assuming certain inputs in terms of formulation composition. In other words, we have been answering questions of type: *If we choose a defined mixture of components A, B and C, what is the cost and what are the expected experimental properties (based on the existing experimental data and the models built on it)?*

Certainly, this is a nice tool, but in most cases the more relevant questions are formulated the other way around, such as: *What is the composition of a mixture with a maximum unit cost of X, for which a certain viscosity of Y is not exceeded and the pull-off force is at least Z?* In addition to cost and scientific considerations, additional constraints may be introduced at later stages of a project, such as the need to stay below a critical level of a certain component to stay within a regulated regime of classification requirements for an end customer product.

In technical terms: We use multiobjective optimization with constraints.

Solution

Use the pymoo-package[1] for leveraging multiobjective optimization in Python.

The idea behind the pyoomo package is to provide as list of constraints for, e. g., a maximum cost and a minimum open time requirement. Based on the available models, the tool will provide a list of possible input configurations (also referred to as *design space*) to achieve the desired output (also referred to as *objective space*) – if feasible – as shown in code section 43.1. To stick with the earlier image of a model as a "bridge" between inputs and outputs (see Figure 37.1 in Concept 37). We will now walk over the bridge from the other direction, i. e., from right to left.

---

1 https://pypi.org/project/pymoo/

https://doi.org/10.1515/9783111334608-051

Listing 43.1: Optimizing for multiple targets under constraints using the pymoo package.

```
1   # %% pymoo tutorial
2   # see https://pymoo.org/getting_started/part_2.html
3
4   import pathlib
5   import numpy as np
6   import pandas as pd
7   import matplotlib.pyplot as plt
8   import plotly.express as px
9
10  from pymoo.core.problem import ElementwiseProblem
11
12  # algorithm
13  from pymoo.algorithms.moo.nsga2 import NSGA2
14  from pymoo.operators.crossover.sbx import SBX
15  from pymoo.operators.mutation.pm import PM
16  from pymoo.operators.sampling.rnd import FloatRandomSampling
17
18  # termination
19  from pymoo.termination import get_termination
20
21  # minimize
22  from pymoo.optimize import minimize
23
24  # compromise programming
25  from pymoo.decomposition.asf import ASF
26
27  from sklearn.utils import check_array
28
29  class MyProblem(ElementwiseProblem):
30
31      def __init__(self):
32          super().__init__(
33              n_var=3, # number of variables/"factors"
34              n_obj=2, # number of objectives/"responses"
35              n_ieq_constr=3,  # number of inequality constraints
36              n_eq_constr=1,  # number of equality constraints
37              xl=np.array([0.0, 0.0, 0.0]),  # lower limits
38              xu=np.array([1.0, 1.0, 1.0])  # upper limits
39              )
40
41          # file to gplearn modeln
42          self.gplearn_models = pd.read_pickle(
43              pathlib.Path().cwd().parent /
44              "073" / "compiled_results.pcl")
45
46          # define name of variables
47          self._names_var = ["A", "B", "C"]
48          self._names_obj = ["time_s", "cost"]
49
50          # info
51          print(self.gplearn_models.iloc[:,:2])
52
53          # info
54          print("MyProblem_initialized")
55
56
57      def _evaluate(self, x, out, *args, **kwargs):
```

```
58          """
59          problem rewritten to "minimize" and "<="
60
61          # variables
62          x0 == A
63          x1 == B
64          x2 == C
65
66          # objective
67          f1 : maximize open time a.k.a. "time_s" == minimize
68              negative(!) "time_s"
69          f2 : minimize cost (defined locally)
70
71          # inequality constraints
72          # see https://pymoo.org/constraints/index.html
73          g1 : A >= 0
74          g2 : B >= 0
75          g2 : C >= 0
76
77          # equality constraint
78          h1 : A + B + C = 1
79
80          walkthrough://vscode_getting_started_page
81          # parameter ranges --> defined in __init__
82          0 .. 1 for all components A, B and C
83
84          """
85
86          # objective functions
87          # model of "time_s"
88          f1 = -self.get_prediction_for(1, x, show_info=False)
89          # model of formulation cost
90          f2 = self.get_cost(x)
91
92          # inequality constraints
93          g1 = 0.25 - x[0]   # min 0.25 of component A
94          g2 = x[1] - 0.4   # maximum B of 0.4
95          g3 = -x[2]
96
97          # # equality constraint(s) ("components have to add up to 1")
98          h1 = x[0] + x[1] + x[2] - 1
99
100         # min/max of objectives
101         out["F"] = [f1, f2]
102         # inequality constraints
103         out["G"] = [g1, g2, g3]
104         # equality constraints
105         out["H"] = [h1]
106
107
108     def get_cost(self, x):
109         # cost function in Euro/kg
110         return x[0]*14.00 + x[1]*8.00 + x[2]*3.75
111
112
113     def get_prediction_for(self, model_nr,
114             experimental_settings, show_info=True):
115
116         # info
117         if show_info:
```

```
118              print("predicting", self.gplearn_models.at[
119                      model_nr, "target_column"])
120              print("with_input_for", self.gplearn_models.at[
121                      model_nr, "factor_columns"])
122
123          # convert to Series
124          experimental_settings = pd.Series({
125                  "A" : experimental_settings[0],
126                  "B" : experimental_settings[1],
127                  "C" : experimental_settings[2]
128                  })
129
130          # get transformer
131          transformer = self.gplearn_models.at[
132              model_nr, "symbolic_transformer"]
133
134          # "harvest" input
135          x_list = []
136          for entry in self.gplearn_models.at[
137                  model_nr, "factor_columns"]:
138              # get input
139              x_list.append(experimental_settings[entry])
140
141          # type conversion
142          X = check_array(np.array([x_list]), dtype=object)
143
144          # transform
145          X_transformed = np.hstack((
146                              X,
147                              transformer.transform(X)
148                              ))
149
150          # get estimator
151          estimator = self.gplearn_models.at[
152              model_nr, "ridge_estimator"]
153
154          # estimate
155          prediction = estimator.predict(X_transformed)[0]
156
157          # return
158          return prediction
159
160
161  # definition of "problem" to be solved
162  problem = MyProblem()
163
164
165  # %% init algorithm
166
167  algorithm = NSGA2(
168          pop_size=1500,   # number of results on the "Pareto Front"
169          n_offspring=5,
170          sampling=FloatRandomSampling(),
171          crossover=SBX(prob=0.9, eta=15),
172          mutation=PM(eta=20),
173          eliminate_duplicates=True
174          )
175
176  # %% define termination criterion
177
```

```
178
179  termination = get_termination("n_gen", 15)
180
181
182  # %% optimize
183
184  res = minimize(
185          problem,
186          algorithm,
187          termination,
188          seed=1,
189          save_history=True,
190          verbose=True
191          )
192
193  # "get" results
194  X = res.X   # --> "design space"
195  F = res.F   # --> "objective space"
196
197  # %% compile predictions
198
199  predictions = pd.concat([
200      pd.DataFrame(X, columns=problem._names_var),
201      pd.DataFrame(F, columns=problem._names_obj)
202      ], axis=1)
203  # correct sign for maximized "time_s" column
204  predictions["time_s"] *= -1
205
206  # show scatterplot of objected space
207  predictions.plot(x="time_s", y="cost", marker="o", linestyle="")
208  plt.show()
209
210  # show design space
211  fig = px.scatter_ternary(predictions, "A", "B", "C",
212                          hover_data=["time_s", "cost"])
213
214  fig.write_html("design_space.html")
```

Note that pymoo allows you to tune numerous parameters and draw different conclusions and recommended actions from the same underlying experimentally supported model, depending on the chosen constraints. In the above code snippet we leveraged *inequality constraints* such as "the relative content of component B must remain below a certain threshold" next to the equality *equality constraint* that "the relative content of all components used must add up to one". Further constraints can be added to the optimization as required.

Increase the values specified as pop_size and in the get_termination function to check the impact on the number of predicted values obtained. You can also try to add additional (in)equality constraints.

Discussion

A basic assumption of the presented approach is the independence of the models. This means that each characteristic, i. e.,output can be described by the corresponding input. In other words: Given a certain set of inputs (in our case merely the relative composition of components A, B and C), we can obtain a prediction for a first property, a prediction for a second property, and so on. The observation that there may be a correlation or even causality between some of the output characteristics (see Concept 44) is not relevant in this scenario.

An alternative package for the purpose of multi-objective optimization is pyomo.[2]

Please note that pymoo is just another tool. Its purpose is to provide a list of options to choose from. In general, pymoo is used to create a link between the *design space* and the *objective space* taking into account the specified constraints. Even though the constraints typically narrow the possibilities dramatically, there may still be numerous "solutions" to choose from. At this point, choosing *your* solution is again a matter of choice/preference.

In the example for introducing the relation between *design space* and *objective space*, a certain constellation of inputs $x1$ and $x2$ is connected to a combination of outputs $f1$ and $f2$ (see Figure 43.1 and Figure 43.2). Due to the assumed constraints, there's a "gap" in both the design and objective space. Interestingly, the "best" solution may move from one edge of the gap to the other depending on the weighting of the importance of satisfying the constraints connected to outputs $f1$ and $f2$.

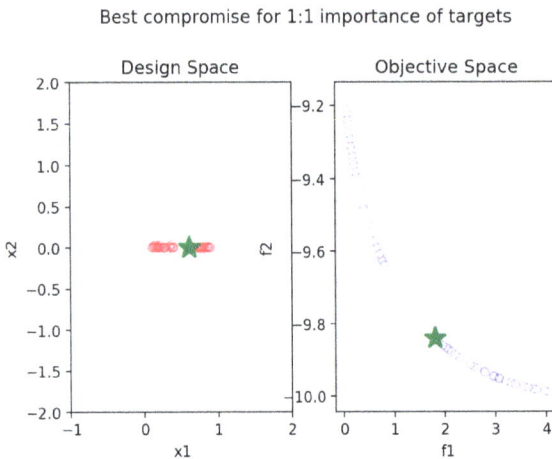

Figure 43.1: Optimal solutions identified by the pymoo-package. Each points corresponds to a solution to the multiobjective optimization problem. The "starred" solution represents the preferred one from this set assuming equal importance of the optimization targets for $f1$ and $f2$.

---

2 https://pypi.org/project/Pyomo/

Best compromise for 10:1 importance of targets

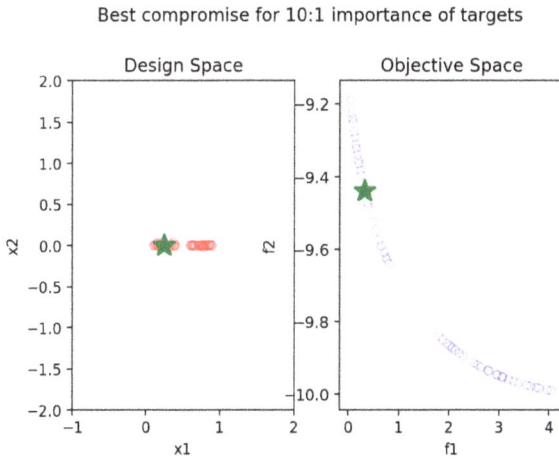

Figure 43.2: Solutions identified by pymoo and highlighted design (starred) assuming a 10:1 weighting of multiobjective optimization goals. In a real-world scenario, this corresponds to, e. g., obtaining recipes for which reducing cost is "more important by a factor of x" compared to meeting a particular experienced property criterion.

44 Ensuring the envisioned causality

Problem

You have a hypothesis regarding the causal relationship between inputs, observed outputs, or just between outputs exclusively that you want to verify / falsify based on existing experimental data.

Context

Having a statistical model is great, understanding causality, i. e.,the *why*, is even what an "understanding" of a particular system is tied to in the mental model of a scientist. This means that even if a perfectly functional "statistical" model is available, a scientist will not be fully satisfied unless he or she understands the *why* of the interactions.

Solution

Use the dowhy-package (read as *do why?*)[1] for conducting causal inference, which supports explicit modeling and testing of causal assumptions.

The idea behind the dowhy-package is to draw cause-and-effect relationships between variables and outcomes. Along with this assumed relationship comes a mental model referred to as *causal graph*, which represents the causes or influences between observables. This graph can also be understood as a hypothesis formulated by a scientist with a certain level of domain knowledge. The dowhy package merely helps to check whether the available data contradicts the specified hypothesis as written down about the causal graph.

The package's easily accessible *four-step interface* aims to encourage causal thinking and analysis in non-experts in the field of machine learning via

1. modeling a causal inference problem taking into account domain knowledge and assumptions, i. e., creating a *causal graph*,
2. identifying an expression for the causal effect considering the assumptions made in the previous step, i. e., the *causal estimand*,
3. estimating the expression using statistical methods, such as matching or instrumental variables, and finally
4. testing the validity of the estimate using a variety of robustness checks.

---

1 https://pypi.org/project/streamlit/

https://doi.org/10.1515/9783111334608-052

In the following example, a set of synthetic data is generated via dowhy and analyzed using the four-step approach introduced above. The data generation and basic visualization is given in code section 44.1.

Listing 44.1: Using dowhy for causal analysis.

```
1   # based on
2   # https://microsoft.github.io/dowhy/example_notebooks/tutorial
3   # -causalinference-machinelearning-using-dowhy-econml.html
4   # https://microsoft.github.io/dowhy/example_notebooks/dowhy_
5   # confounder_example.html?highlight=slope
6
7   # imports
8   import matplotlib.pyplot as plt
9   # import two modules from "dowhy" in one line
10  import dowhy.datasets, dowhy.plotter
11  import seaborn as sns
12
13  # set treatment (viscosity) as causal for the outcome (performance)
14  effect = True
15
16  # build sample data set
17  data_dict = dowhy.datasets.xy_dataset(
18                  20,   # number of samples
19                  effect=effect,
20                  num_common_causes=1,
21                  is_linear=False,
22                  sd_error=0.75
23                  )
24
25  # get DataFrame
26  df = data_dict['df']
27  # rename columns to match the envisioned example
28  df.columns = ["Viscosity", "Performance", "Temperature", "pH"]
29  # tune values
30  df["Temperature"] += 20
31  df["Performance"] += 75
32  # show info / DataFrame structure
33  print(df.head())
34
35  # specify "outcome", "treatment" and "cause"; these are the terms
36  # used in causal analysis
37  outcome = "Performance"
38  treatment = "Viscosity"
39  # one cause
40  common_causes = "Temperature"
41  # # two causes
42  # common_causes = ["Temperature", "pH"]
43
44  # plot sample data via (dowhy built-in) function
45  dowhy.plotter.plot_treatment_outcome(
46          df[treatment],
47          df[outcome],
48          df["pH"]
49          )
50  # show plot
51  plt.show()
52
```

```python
53  # make custom "plot_treatment_outcome"-like plot
54  x = "pH"  # use "pH" as variable on the x axis
55  for _c in [treatment, outcome]:
56      plt.scatter(
57          df[x],
58          df[_c],
59          label=_c
60          )
61  # plot cosmetics
62  plt.xlabel(x)
63  plt.ylabel("\n".join([treatment, outcome]))
64  plt.legend(frameon=False, fontsize=12)
65  # save custom plot
66  plt.savefig(
67          "modified_plotter_type.png",
68          dpi=300,
69          bbox_inches="tight"
70          )
71  # show
72  plt.show()
73
74  # pairplot to show variable dependecies
75  sns.pairplot(
76          df,
77          height=3.5,
78          corner=True,  # don't add axes to the upper triangle
79          diag_kind="hist"
80          )
81
82  # save pairplot
83  plt.show()
84
85  # %% step 1: Model the problem as a causal graph
86  #
87
88  # define model, i.e. build causal graph
89  model= dowhy.CausalModel(
90          data=df,
91          treatment=treatment,
92          outcome=outcome,
93          common_causes=common_causes
94          )
95
96
97  # %% step 2: Identify causal effect using properties of the formal
98  # causal graph
99  #
100
101 identified_estimand = model.identify_effect(
102         proceed_when_unidentifiable=True
103         )
104 print(identified_estimand)
105
106 # %% step 3: Estimate the causal effect
107 #
108
109 estimate = model.estimate_effect(
110         identified_estimand,
111         method_name="backdoor.linear_regression"
112         )
```

```
113
114  print(estimate)
115  # ## Realized estimand
116  # b: Performance~Viscosity+Temperature
117  # Target units: ate
118
119  # ## Estimate
120  # Mean value: 1.876112877792039
121  print(f"DoWhy_estimate_of_causal_effect_is_{estimate.value}")
122  # DoWhy estimate of causal effect is 1.876112877792039
123
124  # Plot slope of line between action and outcome = causal effect
125  dowhy.plotter.plot_causal_effect(
126      estimate,
127      df[treatment],
128      df[outcome]
129      )
130  # show
131  plt.show()
132
133
134  # %% step 3b: Getting more from the "estimate"
135  #
136
137  # get intercept and slope
138  intercept = estimate.intercept
139  slope = estimate.value
140  # get error information
141  std_error = estimate.get_standard_error()[0]
142  ci = estimate.get_confidence_intervals()[0]
143
144  # plot "experimental" data and best causal fit
145  plt.plot(
146          df[treatment],
147          df[outcome],
148          "ko",  # black (k) dot markers (o)
149          zorder=1  # plot markers on top layer
150          )
151  plt.plot(
152          df[treatment],
153          df[treatment]*slope+intercept,
154          "r-",  # red solid line
155          alpha=0.5,  # opacity
156          zorder=1  # plot markers on top layer
157          )
158  # plot fit lines from each experimental point
159  for _i, _row in df[["Viscosity", "Performance"]].iterrows():
160      # info
161      for _s in ci:
162          plt.axline(
163              (_row["Viscosity"], _row["Performance"]),
164              slope=_s,
165              alpha=.10,
166              zorder=0
167              )
168
169  # get axes
170  ax = plt.gca()
171
172  # info text
```

```
173   plt.text(0.05, 0.1,    # x and y-coordinates
174            f"Slope:_{slope:.2f}_$\pm$_{std_error:.2f}",   # 2-decimals
175            transform=ax.transAxes   # use axes coordinate system
176            )
177
178   # specify y-range
179   plt.ylim(bottom=0)
180   # label cosmetics
181   plt.xlabel(treatment)
182   plt.ylabel(outcome)
183
184   # get literal results description
185   estimate.interpret()
186
187   # save plot
188   plt.savefig(
189       "modifed_causal_effect.png",
190       dpi=300,
191       bbox_inches="tight"
192       )
193
194   # show plot
195   plt.show()
196
197   # %% step 4: Refuting the estimate
198   #
199
200   # A) Adding a random common cause variable
201   res_random = model.refute_estimate(
202                   identified_estimand,
203                   estimate,
204                   method_name="random_common_cause"
205                   )
206   print(res_random)
207   # Refute: Add a random common cause
208   # Estimated effect:1.876112877792039
209   # New effect:1.876864110685575
210   # p value:0.49
211
212   # B) Replacing treatment with a random (placebo) variable
213   res_placebo = model.refute_estimate(
214                   identified_estimand,
215                   estimate,
216                   method_name="placebo_treatment_refuter",
217                   placebo_type="permute"
218                   )
219   print(res_placebo)
220   # Refute: Use a Placebo Treatment
221   # Estimated effect:1.876112877792039
222   # New effect:0.025135938683012428
223   # p value:0.48
```

In summary, we have proposed a hypothesis represented by a *causal graph* (see Figure 44.1) and used the rigorous approach provided by dowhy to check whether the so far available (experimental) data contradict this hypothesis. The refutation by the chosen methods confirms the causal effect between viscosity and power in our example (see Figure 44.2).

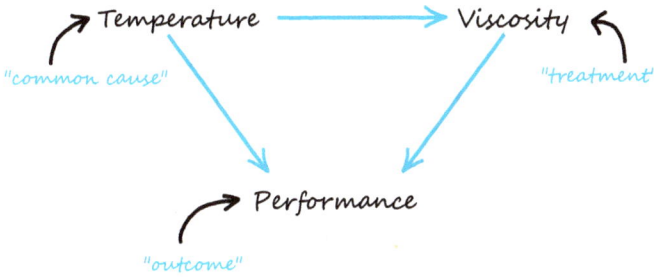

Figure 44.1: Causal graph representing a hypothesis on the relationship between experimental parameters and results; in terms of causal analysis, *Performance* is considered as the *outcome*, *viscosity* as the *treatment* and *temperature* as a *common cause* influencing both viscosity and performance. The question to be answered in the following is whether viscosity (*treatment*) has a causal effect on the observed values of performance (*outcome*), taking into account all available data.

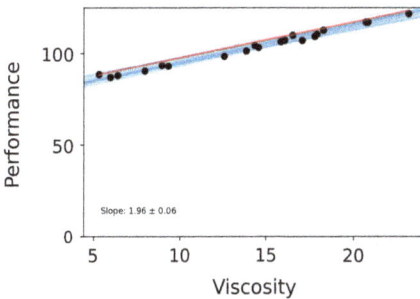

Figure 44.2: Visualization of the causal relationship between performance (outcome on the y-axis) and viscosity (treatment on the x-axis) based on the results of dowhy.

In general, *data* is *not* fitted to a model, but model parameters are adjusted or "tuned" to "fit" the experimental results as represented by the collected *data*. i

---

Discussion

Working with the concept of *causal inference* brings both benefits and challenges.

The previously introduced *causal graph* is to be understood as the underlying (mental) model or hypothesis that is tested "against" the available experimental data. Fortunately, natural scientists create these models regularly in their heads anyway. It is also a step in the overall process of making sense of the relationships within a particular system, where data science methods and expert knowledge or "gut feeling" come together. If you already have some hypotheses, you map them out as a causal graph and test them against your existing data. The challenge associated with these models is that they are

only rarely written down that thoroughly and mostly exist exclusively in the minds of a few individuals. Pinning them down visually has several advantages, including
–   wider and clearer accessibility of the model to colleagues and coworkers fostering higher quality exchange and discussion,
–   the possibility of time-stamped documentation for subsequent visualization of hypothesis evolution, and
–   the possibility for systematic verification of the described hypothesis based on experimental data.

It is crucial to acknowledge that a hypothesis, i. e., causal graph cannot be actively proven correct. Only existing experimental data can be found to be inconsistent with the proposed causal graph. In this sense, the only thing that causal inference can provide is the falsification of a hypothesis described in terms of a causal graph, given (a limited amount of) experimental data. Therefore, if a causal graph has not *yet* been falsified, it could mean that the hypothesized model is actually true, or that you just have not collected sufficient experimental data to falsify it yet. So even given this additional tool, the judgement of experienced scientists is still required.

Further reading

A pioneer in the world of *causal analysis* is Judea Pearl. In his *Book of why* he provides a both informative and entertaining overview of the topic and some reasons why this approach has probably not found its entry into routine natural scientific work. More information in greater detail related to the latest version of dowhy is available online.[2]

---

2 https://github.com/py-why/dowhy

Sharing the project

After you have developed a set of useful functions and tools for your personal daily use, you might consider sharing your progress with colleagues at your institution or simply have an easy option to return to it for later or use another device.

https://doi.org/10.1515/9783111334608-053

45 Building files for distribution

Problem

You have a collection of what you consider to be useful functions and you would like to share them with your colleagues, so that they do not waste their time reproducing existing work.

Context

Let's face it: At some point in your work as a natural scientist you wonder if you are really the first person to try to perform a specific procedure based on an existing experimental file such as parsing its contents for comparison with files of the same type corresponding to other samples. You probably are not. Also, this is something that the vendor of particular piece of equipment should have thought of, not you. In reality, however, in order to be a productive scientist who relies on the foundation of solid data, you'll need to acquire some skills in this area as well. In fact, natural scientific research and development skills seem to be tied to data management and analysis skills more than ever in recent years.

Solution

Use the `poetry-package`[1] to package and distribute your code.

With the appropriate *toml* file at hand (see Concept 12), building the source and wheels archives is as simple as navigating to the project folder and running the

```
poetry build
```

command. Upon completion, a folder named *dist* will be created at the level of the *toml* file, containing the *.tar.gz* and *.whl* files. These can be distributed and used for installation. The files also contain information about the version of the generated package, in our case *0.1.0* (see screenshot Figure 45.1).

---

1 https://pypi.org/project/poetry/

https://doi.org/10.1515/9783111334608-054

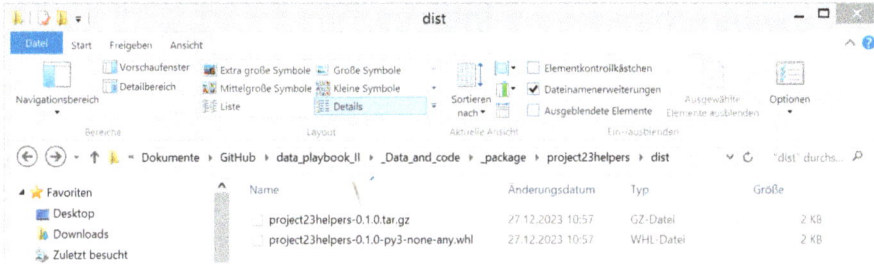

Figure 45.1: Files generated in the *dist* folder by the command `poetry build`.

Discussion

There are numerous ways to package Python projects besides the one suggested by the Python Packaging Authority (PyPA).[2] As outlined in a blog article,[3] available Python packages focus on environment management, Python version management, package management, package building and publishing, or combinations of these. Depending on your focus of work, some tools may be more relevant to your work than others.

Further reading

The generation of files for distribution can be restricted to either *wheel* or *sdist* using the `--format` option of the poetry's `build` command.[4]

---

2 https://packaging.python.org/en/latest/tutorials/packaging-projects/
3 https://alpopkes.com/posts/python/packaging_tools/
4 https://python-poetry.org/docs/cli/#build

46 Pushing to package indices

Problem

You want to share a collection of – considered useful – functions as a Python package via a package index, so that colleagues and/or the Python community do not waste their time reproducing existing work.

Context

Typically, you've been exposed to The Python Package Index (PyPI) as a repository and source of software developed and shared by the Python community early in your learning of Python. Once you feel ready to contribute something to this community, there's an opportunity to change roles from user to package author.

All you need for this, is an account on PyPI.

> **!** To understand the process of publishing a package, it is highly recommended to use TestPyPI instead of PyPI. As the name suggests, it is a separate instance of the The Python Package Index that allows you to try out distribution tools and processes without affecting the real index. Therefore, publishing to TestPyPI will be demonstrated below.

Solution

To publish a package to a package index, we need an account from the appropriate index' site and `poetry` configured accordingly.

Navigate to https://test.pypi.org/, register for an and account and login (see Figure 46.1).

As the next step, we need to obtain a token from TestPyPI for being able to publish the created package in the following. Therefore, we can either follow the link to https://test.pypi.org/manage/account/token/ or navigate there via the *Account Setting* page to get to the page shown in Figure 46.2.

Next, we need to specify in `poetry` where to publish the package and which credentials to use. The repository or *index* to publish to is declared via the `poetry config` command. Here we want to refer to TestPyPI as *test-pypi* and set https://test.pypi.org/legacy/ as the as the Application Programming Interface (API) endpoint, to which the files will be uploaded. Overall, the following command needs to be run to configure `poetry` for publishing to the dedicated endpoint of TestPyPI.

```
1  poetry config repositories.test-pypi https://test.pypi.org/legacy/
```

https://doi.org/10.1515/9783111334608-055

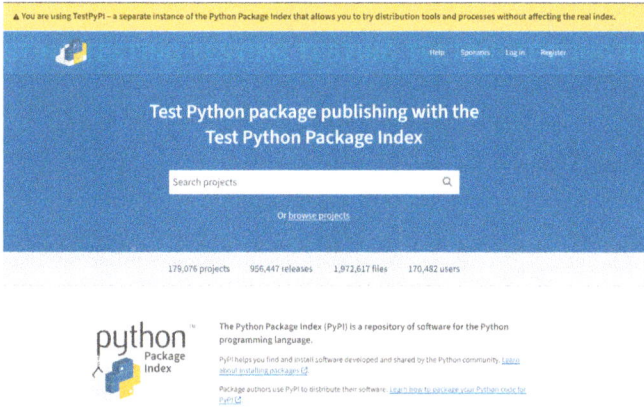

Figure 46.1: Screenshot of the TestPyPI landing page.

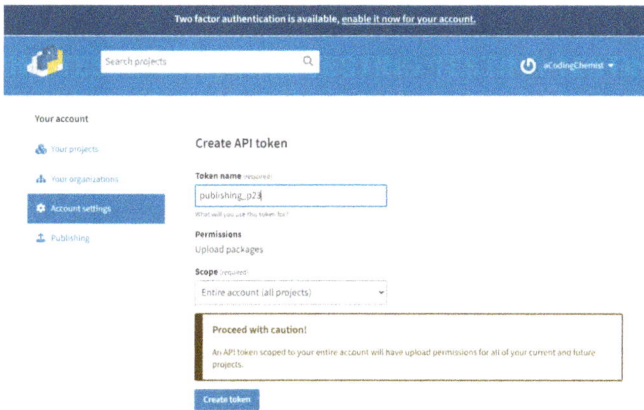

Figure 46.2: Creating a token in TestPyPI.

Once we have specified where we want to publish to, we need to specify how and under which user account we want to do it. The previously obtained token can be connected with the *test-pypi* repository in poetry using the command

```
poetry config pypi-token.test-pypi pypi-XXX
```

where pypi-XXX represents the one-time visible token that is displayed after pressing the *Create token* button shown in Figure 46.2. To finally publish the built package to TestPyPI, we navigate to the directory where the package's *toml* file is located and execute the command

```
1  poetry publish -r test-pypi
```

After a notification in the console, the newly published package is visible in the *Your projects* section of your TestPyPI account.

Discussion

As mentioned before, the purpose of TestPyPI is to familiarize yourself with the process of publishing custom built Python packages to an index. If you decide to publish to the "true" PyPI, the following steps from the above procedure requires some minor adjustments:
- Create and log in to your account at https://pypi.org/.
- Get your token from https://pypi.org/manage/account/token/.
- Store the obtained token in poetry using the command

```
1  poetry config pypi-token.pypi pypi-XXXXXXXX
```

where "pypi-XXXXXXXX" represents the token obtained earlier.
- Publish to PyPI using the command

```
1  poetry publish
```

without explicitly stating the repository to publish to.

! Note that publishing your project to PyPI with a specific version number is necessarily only possible once. This means that after you have made some additions or changes to the source code, you will not be able to republish your package with the same version number. To help with versioning, poetry provides the version command,[1] which shows the current version of the project or "bumps" the version of the project and writes the new version back to *pyproject.toml* if a valid bump rule (such as *major*, *minor* or *patch*) is provided.

Further reading

In addition to using the publicly accessible package indices PyPI and test PyPI, it's also possible to host created packages in private repositories.[2]

---

1 https://python-poetry.org/docs/cli/#version
2 https://packaging.python.org/en/latest/guides/hosting-your-own-index/

47 Sharing `streamlit` applications

Problem

You want to deploy, i. e., make (publicly) available an application developed using `streamlit` (see Concept 41) to share your findings and knowledge with a particular group.

Context

Launching a `streamlit` application via

```
1  streamlit run file.py
```

will allow you to open up the application on your local host in the browser. This might be a big step forward for interfacing and understanding *your* data. However, data and information derived from this representation regularly needs to be understood by larger teams. This holds true for most areas of academic and industrial research.

Solution

Use `streamlit`[1] (see Concept 41) in combination with your GitHub account (see Concept 4) to share your `streamlit` applications based on your (scientific) data.

To start deploying the application, run the application as usual with `streamlit run file.py` (see Concept 41). In the top right corner, you'll find a *Deploy* button as shown in Figure 47.1.

From there, you'll be prompted to choose the *Streamlit Community Cloud* or the *Custom deployment* option according to Figure 47.2.

Continuing with the community cloud option, we are redirected to the https://streamlit.io/ sign in page, where we have the option of logging in with a GitHub account (see Concept 4) as shown in Figure 47.3. After granting access (see Figure 47.4), `streamlit` will be able to deploy an application based on the file(s) residing in an existing GitHub repository from https://share.streamlit.io/.

For deployment, you will need to specify the GitHub repository, branch and files location as shown in Figure 47.5.

After successful deployment, your application will be accessible via the provided Uniform Resource Locator (URL). It's also possible to specify whether the application

---

1 https://pypi.org/project/streamlit/

https://doi.org/10.1515/9783111334608-056

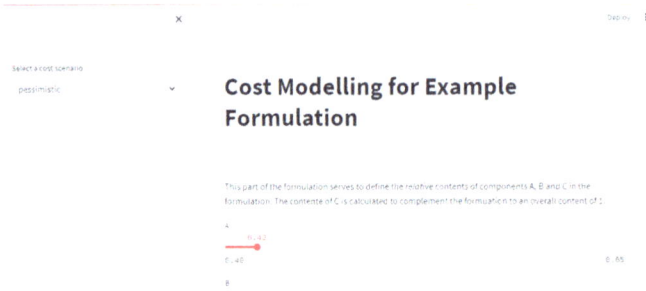

Figure 47.1: Screenshot of the previously (Concept 41) developed `streamlit` application as displayed in the browser.

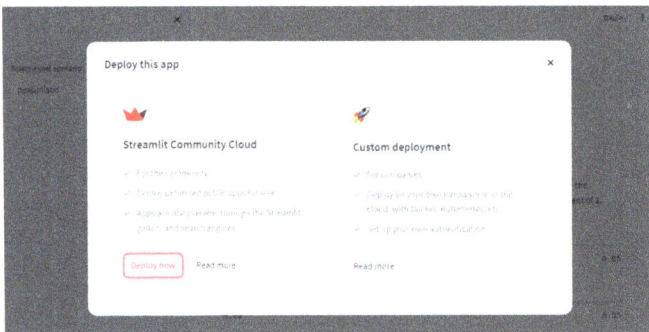

Figure 47.2: Available options after clicking the *Deploy* button shown in Figure 47.1.

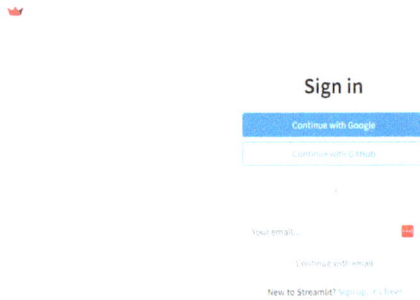

Figure 47.3: Signing in to `streamlit` using a GitHub account.

should be accessible to *anyone* having the link or only to a selection of users on a list identified by their email addresses. Also, the initially assigned URL can be changed in the community version of `streamlit`.[2]

---

2 The above application is available at https://ternaryformulationmodel.streamlit.app/

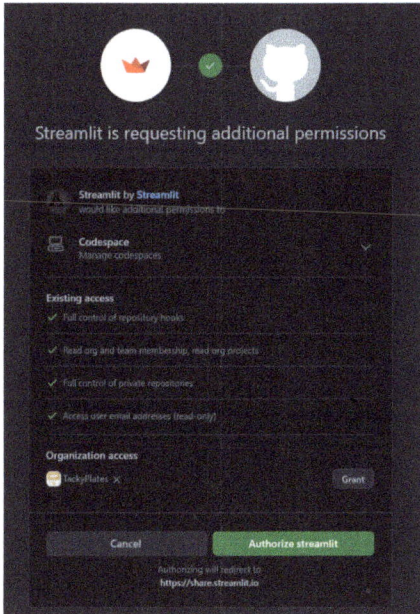

Figure 47.4: Permission request by `streamlit` to GitHub.

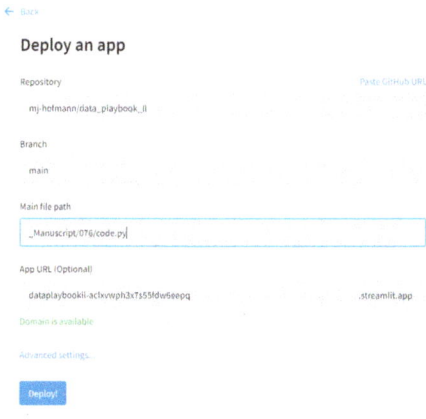

Figure 47.5: For deploying a `streamlit` application in the *Community Cloud*, a GitHub repository, the branch and main file's path within need to be defined.

The application is also available via a browser on your smartphone as shown in Figure 47.6 with a customized Graphical User Interface (GUI).

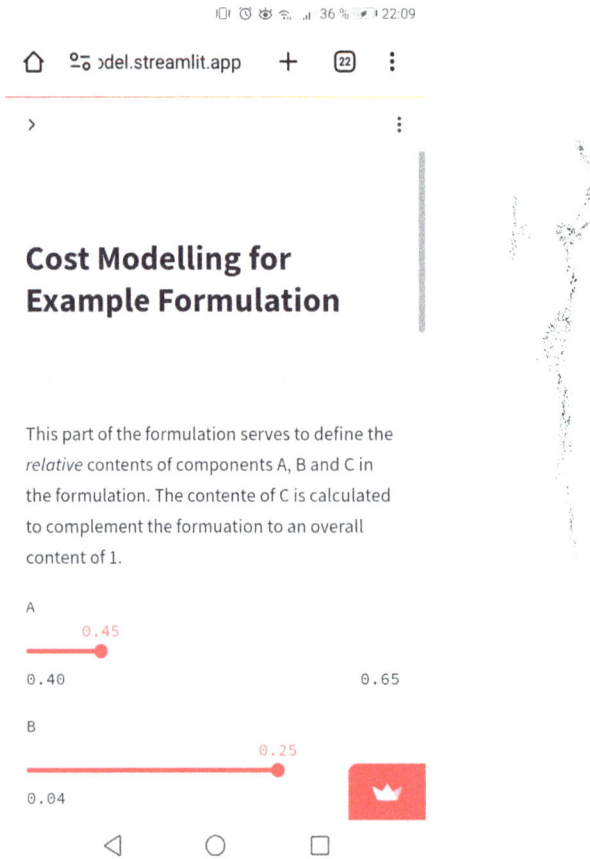

Figure 47.6: The `streamlit` application deployed in the *community cloud* opened on a smartphone.

Discussion

It is crucial to provide a *requirements.txt* in case of relying on additional dependencies.[3]

Further reading

Additional and up-to-date resources related to `streamlit` are available in the `streamlit` blog section https://blog.streamlit.io/, the application gallery, and docs.

---

3 https://docs.streamlit.io/streamlit-community-cloud/deploy-your-app/app-dependencies

Further reading

The following concepts aim to address additional aspects not covered so far. Again, the focus is on making your life as a scientist easier so that you can spend more time on your scientific questions.

https://doi.org/10.1515/9783111334608-057

48 Ensuring code styling via `black`

Problem

You have written a piece of code or script that – finally – does what you want it to do, i. e., the content is correct, but the format does not conform to the many conventions and rules suggested by the Python community.

Context

The coding conventions for the Python code that comprises the standard library in the main Python distribution are summarized in the *PEP8*-document.[1] It also serves as a guide and compilation of best practices for writing Python code.

All too often, people tend to neglect these conventions in the heat of the moment, when the focus is on the content of the code rather than its format.

Solution

Use the `black` package[2] to ensure PEP8-compliant opinionated formatting of your code to "save time and mental energy for more important matters".

After installing `black` using a package management tool such as `pip`, navigate to the directory or file you want to format. In this example, we want to format the code of the file shown in code section 48.1. Therefore, we need to navigate to the directory that contains the file and execute the command

Listing 48.1: Example code *not* formatted with `black`.

```
1  import pathlib
2
3
4  def check_file_type(filename : str):
5      # convert to Path
6      file = pathlib.Path(filename)
7
8      # get extension
9      extension = file.suffix
10
11     # return
12     return  extension
13
14
```

---

1 https://peps.python.org/pep-0008/

2 https://pypi.org/project/black/

https://doi.org/10.1515/9783111334608-058

```
15   # build dict of files to be read
16   files_dict = {
17       "A" : "a.xlsx",
18       "B" : "file_B.pdf",
19       "C": "ResultsFileC.png"
20       }
21
22
23   # dict comprehension
24   file_exts = {k:check_file_type(v) for (k,v) in files_dict.items()}
```

```
1   black code_unformatted.py
```

Please note that formatting with black will not modify the name of the file, only its contents.

After completion of the formatting process, the success message shown in Figure 48.1 is printed to the console.

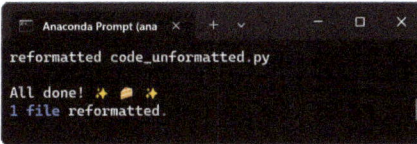

Figure 48.1: Notification of successful code reformatting using black.

As a result, the formatted file (see code section 48.2), no longer contains double subsequent spaces and superficial line breaks, among other things.[3]

Listing 48.2: Code from code section 48.1 formatted using black.

```
1   import pathlib
2
3
4   def check_file_type(filename: str):
5       # convert to Path
6       file = pathlib.Path(filename)
7
8       # get extension
9       extension = file.suffix
10
11      # return
12      return extension
13
```

---

3 An additional line break was manually introduced to meet the formatting requirements of this book.

```
14
15  # build dict of files to be read
16  files_dict = {"A": "a.xlsx", "B": "file_B.pdf",
17               "C": "ResultsFileC.png"}
18
19
20  # dict comprehension
21  file_exts = {k: check_file_type(v) for (k, v) in files_dict.items()}
```

Discussion

The `black` package formats entire files in place. This means that it's not possible to restrict the formatting to certain lines within a file. As mentioned above, this also implies that only the contents of a file are changed, not its name. The same is true for files and folders when the entire directory is passed to `black` for formatting.

Further reading

The code style is fully described online.[4]

---

4 https://black.readthedocs.io/en/stable/the_black_code_style/current_style.html

49 Configuring pre-commit

Problem

You would like to ensure that "housekeeping" tasks related to your codebase, such as code formatting (see Concept 48), are *actually* performed.

Context

When working with a version control system such as git, you would like to identify fundamental problems with the code *before* committing the changes to the repository. Accordingly, the idea is to do all of this in an automated way before actually committing, i. e., the changes.

Solution

Use the `pre-commit` package[1] for managing and maintaining multilingual pre-commit hooks.

As a first step, we will install the pre-commit package using the

```
pip install pre-commit
```

command. To tell `pre-commit` what actions to take before committing, we need to create a configuration file called *.pre-commit-config.yaml* in your repository's root directory. This can be used to point to public repositories. If you are looking for a formatter for Python code, the configuration file could contain the contents shown in code section 49.1.

Listing 49.1: Example *.pre-commit-config.yaml* file for formatting Python code.

```
repos:
-   repo: https://github.com/pre-commit/pre-commit-hooks
    rev: v2.3.0
    hooks:
    -   id: check-yaml
    -   id: end-of-file-fixer
    -   id: trailing-whitespace
-   repo: https://github.com/psf/black
    rev: 22.10.0
    hooks:
    -   id: black
```

---

1 https://pypi.org/project/pre-commit/

https://doi.org/10.1515/9783111334608-059

To install the git hook script, we need to run the command

```
1  pre-commit install
```

which is confirmed by the message *pre-commit installed at .git\hooks\pre-commit* on the command line. With this setup, `pre-commit` will automatically run every time you commit to the repository.

To run `pre-commit` exclusively on a file (such as the previously used unformatted Python file from Concept 48), use the command

```
1  pre-commit run --files code_unformatted.py
```

As this file does not meet the requirements of the formatting options defined in the config file, there will be some "Failed" messages in the console (see Figure 49.1).

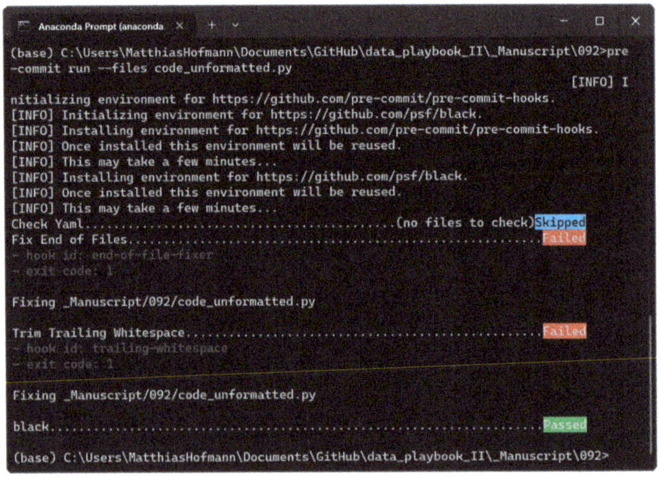

Figure 49.1: Report generated by `pre-commit` after running it against a file *not* meeting the defined formatting requirements.

Running the same command again, will result in an overall "Passed" response (see Figure 49.2), as the detected deviations are taken care of to conform to the best practices and conventions provided by the code formatting tools passed to `pre-commit` via the *.pre-commit-config.yaml* file.

Figure 49.2: Report generated by `pre-commit` after the second pass against a file that meets the defined formatting requirements after the second run.

Discussion

What is the purpose of `pre-commit`? The main idea is to catch perceived "errors", inconsistencies and deviations from generally accepted best practices before you ever commit changes to your repository (see Concept 6). Therefore, the pre-commit hook is leveraged. In general, Git hooks are scripts that run automatically whenever a particular event occurs in a Git repository. They allow you to customize Git's internal behavior and trigger customizable actions at key points in the development lifecycle. They are stored in the hidden *.git* folder inside the main project folder.

This introduced `pre-commit` for formatting Python code, but `pre-commit` works for any programming language. Fortunately, there is already a list of typical pre-commit actions to choose from, such as trimming trailing whitespace to fix the end of a file. A comprehensive list of supported hooks can be found online.[2] If you are missing an action, new hooks can be created from a git repository that is either an installable package or exposes an executable. Therefore, the repository must to contain a *.pre-commit-hooks.yaml* file.[3]

To switch off `pre-commit`, run `pre-commit uninstall`. `i`

Further reading

More information is available at https://pre-commit.com/.

---

2 https://pre-commit.com/hooks.html

3 https://pre-commit.com/#new-hooks

50 Building standalone solutions via PyQt

Problem

You want to work with, i.e., visualize and analyze data within a standalone desktop application with a modern Graphical User Interface (GUI).

Context

So far, we have been mostly working at the script level for reading, analyzing, and displaying scientific data. This might be sufficient for more proficient users, but it limits the application's accessibility to a wider audience.

Solution

Use the PyQt5-package[1] to access Python bindings for the Qt cross-platform C++ libraries. They implement high-level Application Programming Interfaces (APIs) for accessing many aspects of modern desktop and mobile systems. PyQt5 provides all of the elements required to build an attractive GUI. Since we intend to visualize the captured experimental data using plotly (read more about generating interactive plots in Concept 33), we additionally need to install the PyQtWebEngine package.[2]

For working with PyQt5 some experience with Object Oriented Programming (OOP) is certainly beneficial. Here we use the file *main.py* to call our custom window which inherits from the *QMainWindow* class as shown in code section 50.1.

Listing 50.1: Main file for starting the PyQt5 GUI.

```python
from PyQt5.QtWidgets import QApplication

# custom defined main window of the application
import MainWindow

if __name__ == "__main__":
    # create and instanace of QApplication
    app = QApplication([])
    # use the custom MainWindow
    window = MainWindow.MainWindow()
    window.show()
    app.exec_()
```

---

1 https://pypi.org/project/PyQt5/

2 https://pypi.org/project/PyQtWebEngine/

https://doi.org/10.1515/9783111334608-060

The actual functionality, i. e., and its implementation is defined in the *MainWindow.py* file shown in code section 50.2. It defines the size of the window and the title text, as well as the type and positioning of the individual *widgets*. We would like to have an area to display the intended plots, a list of the displayed sample files, a button and a drop-down list to set the plot type. The button serves to toggle between showing experimental raw data, i. e., "curves" and the corresponding extracted characteristics.[3]

Listing 50.2: Implementation of the intended functionality of the PyQt5 GUI.

```
1   from PyQt5 import QtWebEngineWidgets
2   from PyQt5.QtWidgets import QMainWindow
3   from PyQt5.QtWidgets import QWidget, QHBoxLayout
4   from PyQt5.QtWidgets import QListWidget, QPushButton, QComboBox
5   import pathlib
6   import pandas as pd
7   import plotly
8   import plotly.express as px
9
10  import project23helpers.calorimetry as calo
11
12
13  class MainWindow(QMainWindow):
14
15      button_init_text = "Show_Parameters"
```

A screenshot of the resulting GUI is shown in Figure 50.1.

Discussion

An application such as the above standalone GUI based on PyQt5 is where "it all comes together". Alternatively, you can create and share web applications (see Concept 47) to easily distribute your work. Regardless of the type of application chosen, an overarching application will combine many of the previously introduced concepts, from managing the corresponding code with version control, extracting data from experimental results files, packaging these helper packages into your own custom packages, and identifying appropriate visualization methods.

Although a more recent version of PyQt is available at the time of writing,[4] the code example shown above uses PyQt5 as this is part of the used Anaconda distribution.

---

3 In this basic showcase, we limit ourselves to the peak position as the only "characteristic" value. Certainly, there are many more values to consider as characteristics for specific cases.

4 https://pypi.org/project/PyQt6/

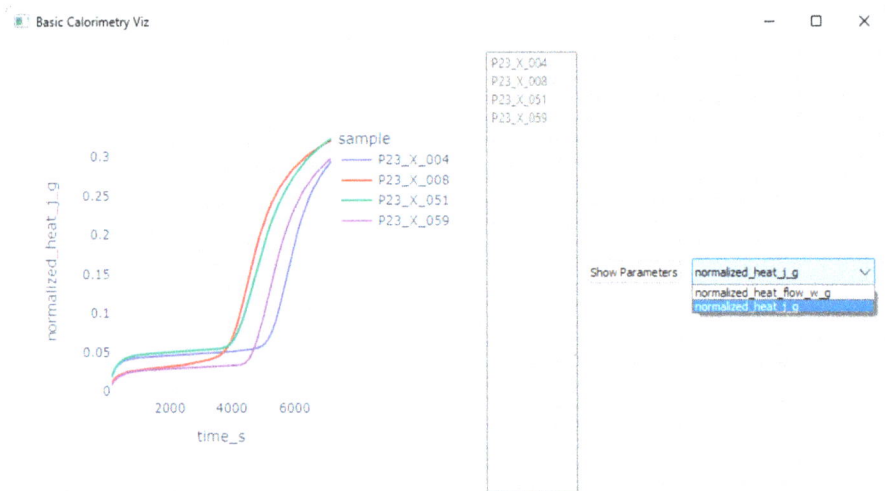

Figure 50.1: Screenshot of the GUI created to visualize heat flow calorimetry data using `PyQt5`.

To convert a `PyQt5`-GUI application into a true standalone solution, additional Python-packages are available, such as `pyinstaller`[5] or `auto-py-to-exe`,[6] which builds on top of the latter.

Further reading

A complete tutorial on creating GUI applications with Python is available online.[7]

---

5 https://pypi.org/project/pyinstaller/

6 https://pypi.org/project/auto-py-to-exe/

7 https://www.pythonguis.com/pyqt5-tutorial/

Concluding remarks

At a high level, natural scientific research and development is a highly collaborative effort, probably more so than ever before. This is especially true in an industrial setting where numerous stakeholders need to be involved.

As a simplified example, consider a system with two design variables x_1 and x_2, and an objective called *response* (see Figure 6). Real-world systems are certainly more complex, but the idea remains the same for higher-dimensional problems.

From a pure *data* perspective (some of) the role's responsibilities are clearly defined. Researchers "set the space" of *where* to look for and *what* to look for. This defines, e. g., which materials should be used to meet a particular performance criterion, but also which methods should be used for quantification.

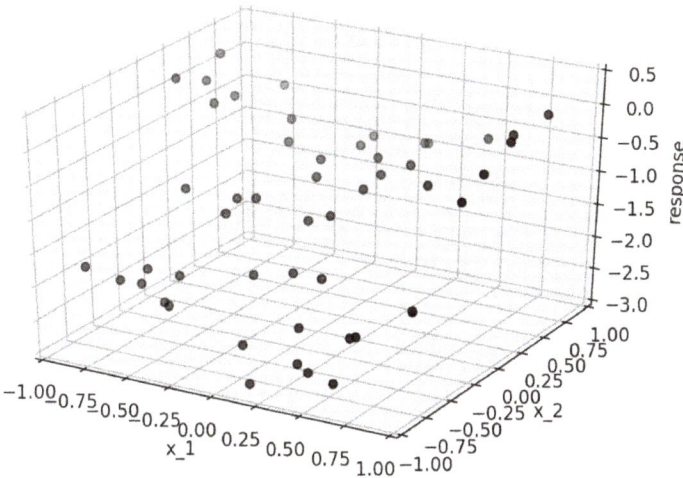

Figure 6: Experimental *response* data collected by scientists and technicians within a design space opened up by the design variables being studied x_1 and x_2.

Next, researchers build a model (either more traditionally in their heads or based on data science methods) to get a broader understanding of the relationships within a studied *system*. This corresponds to generating a theoretical hyperplane in agreement with the so far collected experimental evidence as sketched in Figure 7. Scientists feel they have *understood* a system when they can predict system behavior from limited experimental evidence. A typical question at this stage might be: *What happens to the response if I increase x_1?*

The next hurdle to making a product available to a wider customer base may be regulatory constraints. Due to the increasing awareness and importance of environmental and safety concerns, some components may not be suitable for widespread use in cus-

https://doi.org/10.1515/9783111334608-061

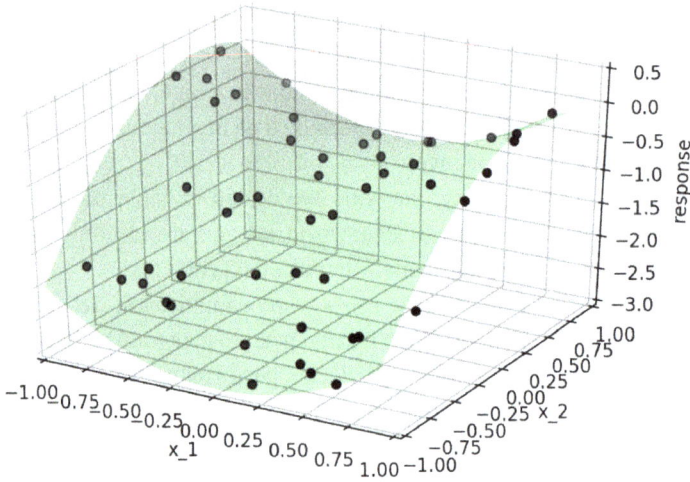

Figure 7: Modeling existing experimental data to obtain an overall picture of the influence of design space components or conditions on the measured result.

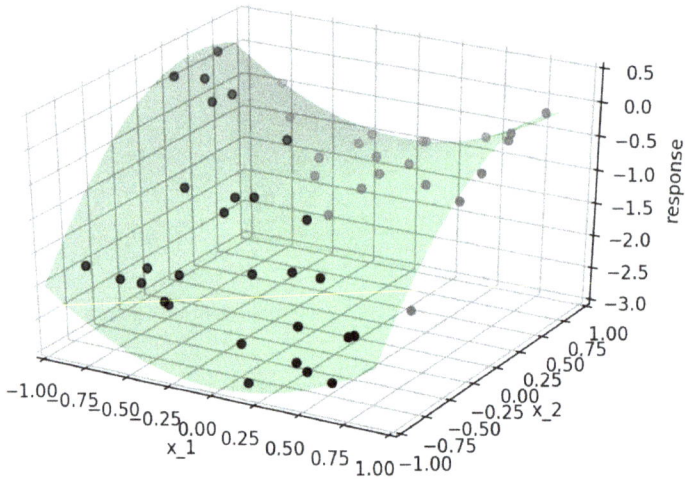

Figure 8: A fraction of possible options are discarded (light grey circles) for regulatory reasons.

tomer applications, even if they offer excellent performance from a technical point of view. Therefore, the introduction of regulatory aspects will crop the accessible design space as shown in Figure 8.

The same principle of "cropping design space" also applies to incorporating additional technical and/or commercial constraints. For example, a certain design space may have to be omitted for legal reasons, or a certain upper cost limit should not be exceeded. Market segmentation also plays an important role. As many products come in a *basic*,

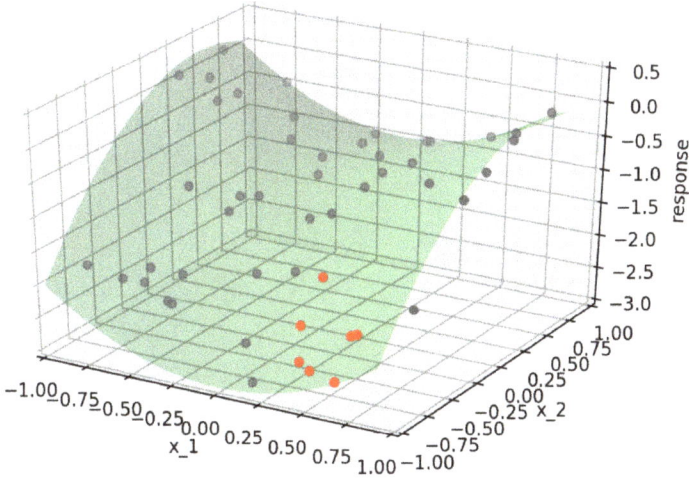

Figure 9: A fraction of possible options being discarded for commercial reasons. The final design space is indicated by red circles.

medium and *premium* version with appropriate price tags, these should be reflected in the technical capabilities. All in all, this leads to further restrictions of the eligible design space for a certain product (see Figure 9).

If the "surviving" options do not fit any additional constraints, there is nothing left to do but loose some of the constraints taken so far or find a clever way to identify additional points on the experimental hyperplane.[1]

Given your creativity and knowledge in the respective field of research and equipped with the tools described in the previous chapters you will be prepared for a new quality of your day-to-day work in advancing your science to new heights.

---

1 Assuming the validity of the created model within limits acceptable to your organization.

List of Figures

https://doi.org/10.1515/9783111334608-062

Index

https://doi.org/10.1515/9783111334608-063

www.ingramcontent.com/pod-product-compliance
Lightning Source LLC
Chambersburg PA
CBHW061412210326
41598CB00035B/6182